Complex numbers

MATTHEW BOULTON COLLEGE

092724

Mathematics for Engineers

The series is designed to provide engineering students in colleges and universities with a mathematical toolkit, each book including the mathematics in an engineering context. Numerous worked examples, and problems with answers, are included.

1. Laplace and z-transforms
2. Ordinary differential equations
3. Complex numbers

Titles in production:

4. Fourier series
5. Differentiation and integration
6. Linear equations and matrices

Mathematics for Engineers

Complex numbers

W. Bolton

Longman
Scientific &
Technical

Longman Scientific & Technical,
Longman Group Limited,
Longman House, Burnt Mill, Harlow,
Essex, CM20 2JE, England
and Associated Companies throughout the world.

© Longman Group Limited 1995

All rights reserved; no part of this publication may be reproduced, stored in a
retrieval system, or transmitted in any form or by any means, electronic,
mechanical, photocopying, recording, or otherwise without either the prior
written permission of the Publishers or a licence permitting restricted copying
in the United Kingdom issued by the Copyright Licensing Agency Ltd,
90 Tottenham Court Road, London, W1P 9HE

First published 1995

British Library Cataloguing in Publication Data
A catalogue entry for this title is available from the British Library

ISBN 0–582–23741–6

Printed in Malaysia

Contents

Preface

This is one of the books in a series designed to provide engineering students in colleges and universities with a mathematical toolkit. In the United Kingdom it is aimed primarily at HNC/HND students and first-year undergraduates. Thus the mathematics assumed is that in BTEC National Certificates and Diplomas, the GNVQ Advanced level or in A level. The pace of development of the mathematics has been aimed at the notional reader for whom mathematics is not their prime interest or "best subject" but need the mathematics in their other studies. The mathematics is developed and applied in an engineering context with large numbers of worked examples and problems, all with answers being supplied.

This book is concerned with complex numbers and their application in, primarily, electrical/electronic engineering. A familarity with basic algebra is assumed. The aim has been to include sufficient worked examples and problems to enable the reader to acquire some understanding and proficiency in the handling of complex numbers and, in particular, their use to describe phasors in electrical circuit analyis.

W. Bolton

1 Introducing complex numbers

1.1 Complex numbers

Complex numbers are a powerful mathematical tool which can be used in many engineering problems. In particular they are used in the analysis of alternating current circuits where phasors are used to represent sinusoidal currents and voltages.

The first three chapters of this book can be considered as a basic introduction to complex numbers, their manipulation and application in alternating current circuit analysis. This chapter is an introduction to the concept of complex numbers and the way they can be represented. Chapter 2 considers the algebraic manipulation of complex numbers, with chapter 3 looking at their use to represent phasors in alternating current circuit analysis.

1.1.1 Roots of a quadratic equation

Consider a quadratic equation

$$x^2 - 4x + 3 = 0$$

The roots of this equation, which is of the form $ax^2 + bx + c = 0$, are given by the equation

$$x = \frac{-b \pm \sqrt{b^2 - 4ac}}{2a} \qquad [1]$$

and are thus

$$x = \frac{+4 \pm \sqrt{16 - 12}}{2}$$

$$x = 2 \pm 1$$

The roots are thus +3 or +1. But now suppose we endeavour to find the roots of the quadratic equation

$$x^2 - 4x + 13 = 0$$

by means of equation [1]. Then

$$x = \frac{+4 \pm \sqrt{16 - 52}}{2}$$

$$x = 2 \pm \sqrt{-9}$$

There is now a problem since we need to find the square root of a negative quantity; squaring $+3$ gives $+9$, hence $\sqrt{+9} = +3$, while squaring -3 gives $+9$, hence $\sqrt{+9} = -3$. There is no real number which has a square of -9.

In order to cope with this problem a number, denoted by j (or sometimes i), is introduced:

$$j = \sqrt{-1}$$

or

$$j^2 = -1$$

Since we cannot obtain a negative number by squaring a real number then the number j is not considered to be a real number and is said to be *imaginary*. Note that

$$j^3 = j^2 j = -\sqrt{-1}$$

$$j^4 = j^2 j^2 = +1$$

The roots $x = 2 \pm \sqrt{-9}$ of the above equation can thus be represented as

$$x = 2 \pm \sqrt{-1(9)} = 2 \pm \left(\sqrt{-1} \right) 3$$

$$= 2 \pm j3$$

The roots are thus made up of two parts, namely a real part and an imaginary part. Such numbers are called *complex numbers*.

In general a complex number z can be represented as

$$z = a + jb \tag{2}$$

where a is the real part of the complex number and b the imaginary part.

Example

Determine the roots of the quadratic equation $2x^2 + 2x + 1 = 0$.

Using the equation [1] for the roots of a quadratic equation, then

$$x = \frac{-2 \pm \sqrt{4-8}}{4} = -0.5 \pm \sqrt{-0.25} = -0.5 \pm j0.5$$

Review problems

1   Determine the roots of the following quadratic equations:

   (a) $x^2 + 25 = 0$, (b) $x^2 + 4x - 5 = 0$, (c) $2x^2 - 2x + 3 = 0$

1.1.2 The complex conjugate

The previous section shows that when a quadratic equation has complex roots that the roots have the form

$$z = a \pm jb \qquad\qquad\qquad [3]$$

There are two roots $a + jb$ and $a - jb$; these two complex numbers being called a *conjugate pair*.
   If we have

$$z = a + jb$$

then the complex conjugate of z, denoted by z^*, is

$$z^* = a - jb$$

Example

What is the complex conjugate of the complex number $5 - j2$?
The complex conjugate is obtained by changing the sign of the imaginary part of the number. Thus the complex conjugate is $5 + j2$.

Review problems

2   What are the complex conjugates of the following complex numbers:

   (a) $-3 + j2$, (b) $5 - j3$, (c) $1 - j0.5$

4   COMPLEX NUMBERS

1.2 Number lines

One way of representing real numbers is to draw a number line, as in figure 1.1. This is a line centred on zero and extending to plus infinity on one side and minus infinity on the other. All real numbers can be represented as points on this line.

Fig. 1.1 The real number line

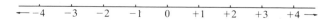

The effect of multiplying a number by a positive constant is to just move the point further out from the origin. For example, multiplying the number +2 by +3 is to move the point from +2 to +6. Multiplying the number −2 by +3 is to move the point from −2 to −6

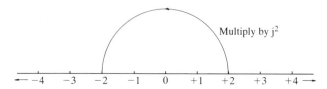

Fig. 1.2 Multiplication by −1

The effect of multiplying a real number by a negative constant, such as −1, is to move the point from one side of the origin to the other, i.e. a rotation of 180° about the origin. Figure 1.2 shows the effect of multiplying the number +2 by −1 to give −2. Since

$$-1 = j^2$$

then we can consider that the real number has been multiplied by j^2 and so

multiplication by $j^2 \equiv 180°$ rotation

If we further multiplied our number −2 by −1, i.e. another j^2, then we end up with + 2. Thus a multiplication of a number + 2 by (−1)(−1) or j^4 is a rotation through a complete 360°, as illustrated in figure 1.3.

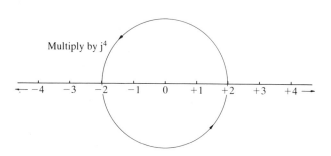

Fig. 1.3 Multiplication by + 1

multiplication by j^4 ≡ 360° rotation

On the basis of the above it would seem logical to equate a rotation through 90° as a multiplication by j and a rotation through 270° as a multiplication by j^3.

multiplication by j ≡ 90° rotation

multiplication by j^3 ≡ 270° rotation

This concept of multiplication by j as involving a rotation is the basis of the use of complex numbers to represent vectors in mechanics or phasors in alternating current circuit analysis (see chapter 3). If we represent a vector or phasor quantity by a line of some length V related to the magnitude of the vector and starting at the origin, then multiplying the vector by j results in a 90° rotation of it, by j^2 a 180° rotation, by j^3 a 270° rotation, by j^4 a 360° rotation (figure 1.4).

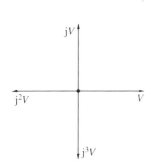

Fig. 1.4 Vectors

1.2.1 The Argand diagram

If we take a real number, say + 2, and multiply it by j, i.e. $\sqrt{-1}$, then the result is equivalent to a rotation through 90° (see previous section). The result is however an imaginary number, for the example chosen j2 or $\sqrt{-4}$. A logical way of thus extending the real number line of figure 1.1 is to consider an entire plane of points, with points having components at right angles to the real number line being imaginary. Figure 1.5 shows such a diagram, it being referred to as an *Argand diagram*.

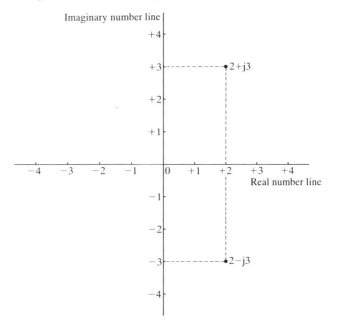

Fig. 1.5 The Argand diagram

The complex numbers 2 + j3 and 2 − j3 have been plotted on the diagram. The complex number 2 + j3 is represented on the diagram by a point arrived at by moving out along the real axis by the amount indicated by the real number, namely 2, and then rotating through 90°, since we have j, and moving out a distance of 3. The complex number 2 − j3 is represented on the diagram by a point arrived at by moving out along the real axis by 2 and then rotating through 270°, since −j = j³, and moving out a distance of 3. This is just the same as plotting *x* and *y* co-ordinates of a point with the real part of the complex number being the *x* co-ordinate and the imaginary part the *y* co-ordinate. The numbers plotted on the Argand diagram are complex conjugates. If you think of the real number axis being a plane mirror, then one of the conjugate pair can be considered to be the reflection of the other one, being like an image and as far behind the mirror as the other one is in front.

Review problems

3   Plot on an Argand diagram the following complex numbers:

   (a) −2 + j3, (b) 2 − j1, (c) −3 − j2, (d) 1 + j2

1.3 Cartesian and polar forms

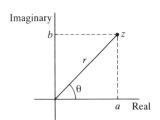

Fig. 1.6 The complex number *z*

Complex numbers written as $z = a + jb$ are said to be in the *Cartesian form*. This is because *a* and *b* specify the co-ordinates on the Argand diagram (figure 1.6).

However, a point on an Argand diagram can also be specified by the length *r* of the line joining it to the origin and the angle θ it makes with the positive *x*-axis, when rotating in an anticlockwise direction (figure 1.6). The length *r* is called the *modulus* of the complex number and written as |z|. The angle θ is called the *argument* of the complex number and written as arg *z*. Thus we can specify a complex number by giving the values of *r* and θ. This can be done by writing the complex number in the form

$$z = r\angle\theta \qquad [4]$$

This is called the *polar form* of the complex number.

From figure 1.6 it can be seen that, by the Pythagoras theorem,

$$|z| = r = \sqrt{(a^2 + b^2)} \qquad [5]$$

and

$$\arg z = \theta = \tan^{-1}\left(\frac{b}{a}\right) \qquad [6]$$

Thus the complex number $z = 2 + j4$ can be represented on the Argand diagram by a point which is at the end of a line, starting at the origin, of length

$$|z| = r = \sqrt{2^2 + 4^2} = 4.5$$

This would make an angle to the positive real number axis of

$$\text{arg} = \theta = \tan^{-1}\left(\frac{4}{2}\right) = 63.4°$$

Thus in polar form the complex number is $4.4\angle 63.4°$.

Consideration of figure 1.6, which shows the complex number $z = a + jb$ on an Argand diagram, indicates that

$$a = r\cos\theta \qquad\qquad\qquad [7]$$

and

$$b = r\sin\theta \qquad\qquad\qquad [8]$$

Hence we can write

$$z = r\cos\theta + jr\sin\theta$$

and so

$$z = r\angle\theta = r(\cos\theta + j\sin\theta) \qquad\qquad\qquad [9]$$

Note that when the angle is more than simply just a quantity like 30°, brackets might be used, e.g. $\angle(-30°)$. Alternatively, this can be written as $\underline{/-30°}$. The intention is to clearly indicate what is encompassed by the angle term.

Example

Determine the modulus and argument of the complex numbers
(a) $2 + j2$, (b) $-1 + j2$, (c) $-2 - j2$.

Figure 1.7 shows the points plotted on Argand diagrams. It is often worthwhile doing this in order see what quadrant θ is in.
For (a)

$$|z| = \sqrt{2^2 + 2^2} = 2.8$$

$$\text{arg} = \theta = \tan^{-1}\left(\frac{2}{2}\right) = 45°$$

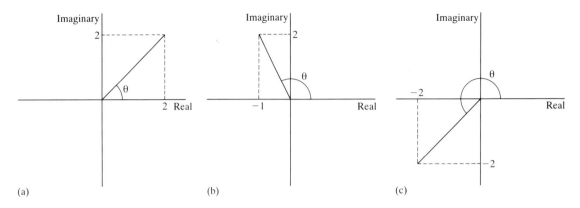

(a)   (b)   (c)

Fig. 1.7 Example

In polar form the number is thus $2.8\angle45°$.

For (b)

$$|z| = \sqrt{\left[(-1)^2 + 2^2\right]} = 2.2$$

The angle is in the second quadrant and is thus between 90° and 180°. Such an angle will have a negative tangent. Thus

$$\arg = \theta = \tan^{-1}\left(\frac{2}{-1}\right) = 116.6°$$

In polar form the number is $2.2\angle116.6°$.

For (c)

$$|z| = \sqrt{\left[(-2)^2 + (-2)^2\right]} = 2.8$$

The angle is in the third quadrant and is thus between 180° and 270°. Such an angle will have a positive tangent. Thus

$$\arg = \theta = \tan^{-1}\left(\frac{-2}{-2}\right) = 225°$$

In polar form the number is $2.8\angle225°$.

Example

Write the complex number $10\angle60°$ in Cartesian form.

Using equation [7],

$$a = 10\cos60° = 5.0$$

Using equation [8],

$$b = 10 \sin 60° = 8.7$$

Hence the complex number in Cartesian form is $5.0 + j8.7$.

Example

Write the complex number $-2 - j2$ in polar form.

Figure 1.8 shows this number on an Argand diagram.

$$r = |z| = \sqrt{\left[(-2)^2 + (-2)^2\right]} = 2.8$$

The angle will be in the third quadrant and so will lie between $180°$ and $270°$.

$$\theta = \tan^{-1}\left(\frac{-2}{-2}\right) = 225°$$

Hence the number in polar form is $2.8\angle225°$. When the angle is positive then the angle is considered to be, as indicated in the diagram, obtained by rotating anticlockwise from the positive real number axis. The complex number may also be written as $2.8\angle(-135°)$. The minus sign indicates that the angle is reckoned by rotating clockwise from the positive real number axis.

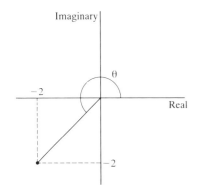

Fig. 1.8 Example

Review problems

4 Determine the modulus and argument of :
   (a) $3 + j2$, (b) $-1 + j1$, (c) $1 - j1$.
5 Express the following in polar form:
   (a) $3 + j4$, (b) $-j4$, (c) 3, (d) $-5 + j8$, (e) $5 - j8$.
6 Express the following in Cartesian form:
   (a) $4\angle30°$, (b) $6\angle180°$, (c) $2\angle0°$, (d) $7\angle(-145°)$.

1.4 Vectors and phasors

Consider how we can describe a vector quantity by complex numbers. A vector quantity is a quantity that has both a magnitude and a direction. It can be represented by an arrow-headed line with the length of the line representing the magnitude and the angle it makes with some reference axis the direction of the vector. Suppose we have a vector of magnitude 1 and in the direction of the reference axis as the line shown on figure 1.9. We will call this the *unit reference vector*. If we have another vector of magnitude 1 but at some other angle then we can specify it in terms of the unit reference vector and how much it has to be rotated to reach this angle. If a vector has a different magnitude then all we have to

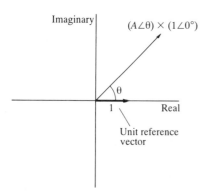

Fig. 1.9 Representing a vector

do is to state by what factor the magnitude of the unit reference vector has to be multiplied to give the required length. Thus, if we use the polar notion of complex numbers we would specify the unit reference vector as $1\angle 0°$ and a vector of magnitude A at an angle θ to the reference axis as

$$(A \times 1)\angle(\theta + 0°) = (A\angle\theta) \times (1\angle 0°)$$

Generally the existence of the unit reference vector is taken as understood and so a vector of magnitude A at an angle θ to the reference axis is just represented by $A\angle\theta$.

For example, suppose we have a vector of magnitude 10 N which is at an angle of 30° to the reference axis. In polar notion we can represent the phasor as $10\angle 30°$ N. Alternatively we can represent the phasor in Cartesian notion, using equation [9], as

$$10\angle 30° = 10(\cos 30° + j \sin 30°) = 8.7 + j5.0 \text{ N}$$

Phasors are used to specify voltages and currents which vary sinusoidally with time. They are rather like vectors. In the case of phasors the unit reference phasor is one of length 1 at an angle to the reference axis of 0°. It thus represents a sinusoidally varying quantity of unit amplitude which has zero angle at time $t = 0$, i.e. a sinusoid which starts with the value 0 at $t = 0$. All other phasors are then expressed, like the vectors, in terms of their phase angle, relative to the reference phasor at time $t = 0$ and their magnitude is terms of how much bigger the maximum or root-mean-square value of the voltage or current is than that of the unit reference phasor. See chapter 3 for more discussion of phasors.

Review problems

7 Represent the following vectors by both polar and Cartesian complex notation:

(a) a force of 10 N at an angle of 60° to the reference axis,
(b) a force of 1 kN at an angle of 120° to the reference axis,
(c) a velocity of 5 m/s in a direction at 45° to the reference axis,
(d) a velocity of 10 m/s in a direction of 140° to the reference axis.

Further problems

8 Determine the roots of the following equations:
(a) $x^2 + 16 = 0$, (b) $2x^2 + 2x + 5 = 0$, (c) $x^2 - 9 = 0$,
(d) $2x^2 - x + 1 = 0$, (e) $x^2 + 2x + 6 = 0$.

9 What are the complex conjugates of the following complex numbers?
(a) $-2 + j3$, (b) $5 + j6$, (c) $-7 - j7$.

10 Express the following complex numbers in polar form:
(a) $-1 + j3$, (b) $-1 - j3$, (c) $1 + j3$, (d) 1, (e) $j3$.

11 Express the following in Cartesian form:
(a) $10\angle0°$, (b) $10\angle90°$, (c) $10\angle(-90°)$, (d) $10\angle30°$,
(e) $10\angle120°$.

12 Represent the following vectors by both polar and Cartesian complex notation:
(a) a force of 100 N at an angle of 30° to a reference axis,
(b) a force of 100 N at an angle of 90° to a reference axis,
(c) a velocity of 12 m/s in a direction of 30° to a reference axis,
(d) a velocity of 4 m/s in a direction of 180° to a reference axis,
(e) an acceleration of 10 m/s^2 in a direction of 45° to a reference axis,
(f) an acceleration of 5 m/s^2 in a direction of 230° to a reference axis.

2 Complex number algebra

2.1 Operations with complex numbers

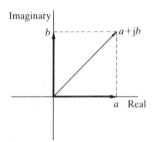

Fig. 2.1 A complex number as a vector quantity

On an Argand diagram, a complex number $a + jb$ can be considered to be, by means of the parallelogram of vectors, the vector sum of one vector of length a along the real axis and one of length b along the imaginary axis. Figure 2.1 illustrates this. Thus we can think of the complex number $a + jb$ as representing a vector quantity. Written in polar notation as $|z|\angle\theta$, then its representation as a vector of length $|z|$ at an angle θ is evident.

Two complex numbers are equal only if their real parts are equal and their imaginary parts are equal. Thus $2 + j3$ is *not* equal to $3 + j2$. If these complex numbers are written in polar notation then we have $3.6\angle56.3°$ and $3.6\angle33.7°$ and it is clear that they are not equal. If these two numbers are plotted on an Argand diagram then it is clear that they represent two different points. Similarly $2 + j3$ and $2 - j3$ are *not* equal, just as $2 + j3$ and $-2 + j3$ are *not* equal.

Two complex numbers are conjugates (see section 1.1.2) if their real parts are equal and the imaginary parts are positive in one case and minus the same number in the other. Thus $2 + j3$ and $2 - j3$ are complex conjugates. On an Argand diagram, if you imagine the real number axis to be a plane mirror then the complex conjugates are mirror images of each other.

The operations of addition, subtraction, multiplication and division can all be performed on complex numbers. The Cartesian and polar forms of complex numbers each have their own merits for different forms of computation. This chapter discusses these operations, with chapter 3 providing illustrations of their use in alternating current circuit analysis with phasors.

Example

What must be the values of a and b for the two complex numbers $a + jb$ and $5 + j2$ to be equal?

To be equal the real parts must be equal and the imaginary parts must be equal. Thus we must have $a = 5$ and $b = 2$.

Example

What is the complex conjugate of $5 + j2$?

The complex conjugate will have the same real number but the imaginary number will be of the opposite sign. Thus the complex conjugate is $5 - j2$.

Review problems

1   What values of a and b are required to make the complex numbers $a + jb$ and $2 + j7$ equal?
2   What are the complex conjugates of (a) $3 + j6$, (b) $1 - j2$?

2.2 Addition and subtraction

To add complex numbers we simply add the real parts and add the imaginary parts.

$$(a + jb) + (c + jd) = (a + c) + j(b + d) \qquad [1]$$

Thus $2 + j3$ plus $3 + j5$ is $5 + j8$.

Consider what is happening with addition in terms of an Argand diagram (figure 2.2). Adding the two complex numbers is the same as the vector addition of two vectors, using the parallelogram of vectors. This method of adding complex numbers in the Cartesian form is much simpler than if they were in polar form and had to add them in the ways traditionally used for the addition of vector quantities by the parallelogram of vectors, i.e. by a scale drawing or calculation based on the triangle formed by the vectors and the resultant.

To subtract complex numbers we simply subtract the real parts and subtract the imaginary parts.

$$(a + jb) - (c + jd) = (a - c) + j(b - jd) \qquad [2]$$

Thus $3 + j4$ minus $2 + j6$ is $1 - j2$.

Consider what is happening with subtraction in terms of an Argand diagram (figure 2.3). To subtract two vector quantities the method used is for the vector being subtracted to be turned around to the opposite direction and then this vector added. Thus turning the vector $c + jd$ around gives $-c - jd$. Adding this complex number to $a + jb$ gives the result shown in equation [2] above.

Addition and subtraction is simplest with complex numbers in the Cartesian form.

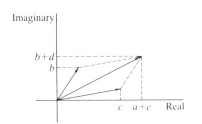

Fig. 2.2 Adding complex numbers

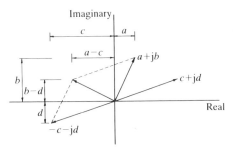

Fig. 2.3 Subtracting complex numbers

Example

Add the two complex numbers $4 + j3$ and $5 - j7$.

Adding the real parts gives 9, adding the imaginary parts gives -4. Thus the sum is $9 - j4$.

Example

Subtract $2 - j4$ from $5 + j2$.

Subtracting the real parts gives $5 - 2 = 3$, subtracting the imaginary parts gives $2 - (-4) = 6$. Hence the result is $3 + j6$.

Review problems

3 Obtain (i) $z_1 + z_2$, (ii) $z_1 - z_2$, for the following complex numbers:
(a) $z_1 = 1 + j5$, $z_2 = 3 + j2$,
(b) $z_1 = 3 - j4$, $z_2 = 1 + j6$,
(c) $z_1 = -2 + j3$, $z_2 = 3 + j4$,
(d) $z_1 = 3 + j4$, $z_2 = -1 - j2$.

2.3 Multiplication

Consider the multiplication of two complex numbers expressed in the Cartesian form, namely $a + jb$ and $c + jd$.

$$z = (a + jb)(c + jd)$$

$$= ac + j(ad + bc) + j^2bd$$

However, $j^2 = -1$ and so

$$z = ac + j(ad + bc) - bd \qquad [3]$$

Thus if, for example, we multiply $2 + j3$ and $4 + j2$ then we have

$$z = (2 + j3)(4 + j2) = 8 + j16 - 6 = 2 + j16$$

Now consider multiplication of two complex numbers when they are in polar form, i.e. $z_1 = |z_1|\angle\theta$ and $z_2 = |z_2|\angle\phi$. We can write these numbers as (equation [9], chapter 1)

$$z_1 = |z_1|(\cos\theta + j\sin\theta)$$

$$z_2 = |z_2|(\cos\phi + j\sin\phi)$$

Thus

$$z = |z_1 z_2|[\cos\theta\cos\phi + j(\sin\theta\cos\phi + \cos\theta\sin\phi) \\ +j^2 \sin\theta\sin\phi]$$

However, $j^2 = -1$ and since $\cos\theta\cos\phi - \sin\theta\sin\phi = \cos(\theta+\phi)$ and $\sin\theta\cos\phi + \cos\theta\sin\phi = \sin(\theta+\phi)$ then

$$z = |z_1 z_2|[\cos(\theta+\phi) + j\sin(\theta+\phi)]$$

$$z = |z_1 z_2|\angle(\theta+\phi) \qquad\qquad [4]$$

Multiplication of two magnitudes gives a resultant with a magnitude which is the product of the two magnitudes, i.e. $z = |z_1||z_2|$. If we multiply a vector by an angle all we are doing is just rotating that vector by that angle, the resultant angle is the sum of the two separate angles. Thus, for example, the product of the two complex numbers $2\angle20°$ and $3\angle40°$ is $6\angle60°$.

Example

What is the product of the complex numbers $5 + j2$ and $3 - j4$?

The product is given by

$$z = (5 + j2)(3 - j4) = 15 + j(6 - 20) - j^2 8 = 23 - j14$$

Example

What is the product of the complex numbers $5\angle50°$ and $2\angle60°$?

The product is given by

$$z = |5 \times 2|\angle(50° + 60°) = 10\angle110°$$

Example

What is the product of the two complex numbers $3\angle60°$ and $4\angle(-20°)$?

The product is given by

$$z = |3 \times 4|\angle(60° - 20°) = 12\angle40°$$

Review problems

4 Determine the products of the following complex numbers:
   (a) $2 + j5$ and $1 - j2$, (b) $3 + j5$ and $-2 + j3$,
   (c) $-2 + j2$ and $4 + j3$, (d) $1 + j2$ and $5 - j2$.

5   Determine the products of the following complex numbers:
(a) $2\angle 60°$ and $5\angle 10°$, (b) $3\angle 40°$ and $5\angle 120°$,
(c) $6\angle(-20°)$ and $2\angle 30°$, (d) $2\angle 40°$ and $3\angle(-60°)$.

2.2.1 Multiplication of conjugate numbers

Consider the multiplication of a complex number $a + jb$ by its conjugate $a - jb$.

$$z = (a + jb)(a - jb) = a^2 + b^2 \qquad [5]$$

Whenever a complex number is multiplied by its conjugate then the result is a real number.
   Thus if we have the complex number $2 + j3$ and we want to convert it into a real number, we can multiply it by its conjugate of $2 - j3$. The result is $2^2 + 3^2 = 13$.

Review problems

6   Determine the result of multiplying the following complex numbers by their conjugates:
(a) $4 + j3$, (b) $2 - j5$, (c) $-2 + j6$, (d) $4 - j5$.

2.4 Division

Consider the division of $a + jb$ by $c + jd$.

$$z = \frac{a + jb}{c + jd}$$

To carry out this division it is necessary to make the denominator a real number. We can do this by multiplying it by its conjugate. Thus we have

$$z = \frac{a + jb}{c + jd} \times \frac{c - jd}{c - jd} = \frac{(a + jb)(c - jd)}{c^2 + d^2} \qquad [6]$$

Consider as an example the division of $2 + j3$ by $4 + j2$.

$$z = \frac{2 + j3}{4 + j2} \times \frac{4 - j2}{4 - j2} = \frac{8 + j8 - j^2 6}{4^4 + 2^2} = \frac{14 + j8}{4^2 + 2^2} = 0.7 + j0.4$$

   Now consider the division of the two complex numbers in polar form $z_1 = |z_1|\angle\theta$ and $z_2 = |z_2|\angle\phi$.

$$z = \frac{|z_1|(\cos\theta + j\sin\theta)}{|z_2|(\cos\phi + j\sin\phi)}$$

We can make the denominator a real number by multiplying it by its conjugate. Thus

$$z = \frac{|z_1|(\cos\theta + j\sin\theta)}{|z_2|(\cos\phi + j\sin\phi)} \times \frac{\cos\phi - j\sin\phi}{\cos\phi - j\sin\phi}$$

$$= \frac{|z_1|(\cos\theta + j\sin\theta)(\cos\phi - j\sin\phi)}{|z_2|(\cos^2\phi + \sin^2\phi)}$$

However, $\cos^2\phi + \sin^2\phi = 1$ and so

$$z = \frac{|z_1|}{|z_2|}[(\cos\theta\cos\phi + \sin\theta\sin\phi) + j(\sin\theta\cos\phi - \cos\theta\sin\phi)]$$

Since the first bracketed term is the cosine of the difference of the two angles and the second term the sine of the difference of the two angles,

$$z = \frac{|z_1|}{|z_2|}[\cos(\theta - \phi) + j\sin(\theta - \phi)]$$

$$= \frac{|z_1|}{|z_2|}\angle(\theta - \phi) \qquad\qquad [7]$$

Thus to divide the two complex numbers we divide their magnitudes and subtract their angles.

Consider the division of $4\angle 40°$ by $5\angle 20°$. Division of the magnitudes gives $4/5 = 0.8$. Subtraction of the angles of the complex numbers gives $40° - 20° = 20°$. Hence the result is $0.8\angle 20°$.

Example

Determine, in Cartesian form, the value of $2 + j3$ divided by $4 + j2$.

$$z = \frac{2 + j3}{4 + j2} \times \frac{4 - j2}{4 - j2} = \frac{8 + j8 - j^2 6}{4^2 + 2^2} = 0.7 + j0.4$$

Example

Determine, in polar form, the value of $10\angle 40°$ divided by $4\angle(-10°)$.

$$z = \frac{10\angle 40°}{4\angle(-10°)} = \frac{10}{4}\angle\{40° - (-10°)\} = 2.5\angle 50°$$

Review problems

7   Determine, in Cartesian form, z_1/z_2 for the following:
   (a) $z_1 = 4 + j2$, $z_2 = 1 + j4$,
   (b) $z_1 = 3 + j2$, $z_2 = 2 - j1$,
   (c) $z_1 = 4 - j2$, $z_2 = 2 + j2$,
   (d) $z_1 = -2 + j3$, $z_2 = 2 - j5$.

8   Determine, in polar form, z_1/z_2 for the following:
   (a) $z_1 = 4\angle 40°$, $z_2 = 2\angle 20°$,
   (b) $z_1 = 6\angle 30°$, $z_2 = 3\angle(-40°)$,
   (c) $z_1 = 4\angle(-20°)$, $z_2 = 6\angle 40°$,
   (d) $z_1 = 8\angle 50°$, $z_2 = 10\angle(-10°)$.

2.5 Evaluating complex equations

The basic rules for adding, subtracting, multiplying and dividing complex numbers have been given earlier in this chapter. From them we can deduce such relationships as:

$$z_1 + z_2 = z_2 + z_1$$

$$(z_1 + z_2) + z_3 = z_1 + (z_2 + z_3)$$

$$z_1 z_2 = z_2 z_1$$

$$(z_1 z_2) z_3 = z_1 (z_2 z_3)$$

$$z_1 (z_2 + z_3) = z_1 z_2 + z_1 z_3$$

and for polar notation:

$$\frac{1}{|z|\angle\theta} = \frac{1}{|z|}\angle(-\theta)$$

$$(|z|\angle\theta)^2 = |z|^2\angle 2\theta$$

$$(|z|\angle\theta)^3 = |z|^3\angle 3\theta$$

In general, adding and subtracting is easiest with complex numbers in Cartesian form, multiplication and division easiest when they are in polar form.

Example

With $z_1 = 2 - j3$ and $z_2 = 1 + j2$, determine:
(a) $z_1 + z_2$, (b) $z_1 - z_2$, (c) z_1/z_2, (d) $(1/z_1) + (1/z_2)$.

(a) Following the procedures for the addition of complex numbers in Cartesian form,

$$z_1 + z_2 = 2 - j3 + 1 + j2 = 3 - j1$$

(b) Following the procedures for the subtraction of complex numbers in Cartesian form,

$$z_1 - z_2 = (2 - j3) - (1 + j2) = 1 - j5$$

(c) Following the procedures for the division of complex numbers in Cartesian form,

$$\frac{z_1}{z_2} = \frac{2 - j3}{1 + j2} \times \frac{1 - j2}{1 - j2} = \frac{-4 - j7}{1^2 + 2^2} = -0.8 - j1.4$$

(d) Each part of this problem can be considered as a separate division, before the results are then combined.

$$\frac{1}{z_1} + \frac{1}{z_2} = \frac{1}{2 - j3} \times \frac{2 + j3}{2 + j3} + \frac{1}{1 + j2} \times \frac{1 - j2}{1 - j2}$$

$$= \frac{2 + j3}{13} + \frac{1 - j2}{5}$$

$$= \frac{10 + j15 + 13 - j26}{65}$$

$$= 0.35 - j0.17$$

Example

Simplify the following, giving the answer in polar form:

$$\frac{(4\angle 20°)^2 \times (2\angle 40°)}{8\angle 30°}$$

Since this involves just multiplication and division, the problem can be tackled with the quantities left in polar form. Thus

$$\frac{(4\angle 20°)^2 \times (2\angle 40°)}{8\angle 30°} = \frac{(16\angle 40°) \times (2\angle 40°)}{8\angle 30°}$$

$$= \left| \frac{16 \times 2}{8} \right| \angle(40° + 40° - 30°) = |4|\angle 50°$$

Example

Simplify $2\angle 20° + 3\angle 40°$, giving the answer in polar form.

The addition of two complex quantities is simplest if they are in Cartesian form. Hence the problem is tackled as follows:

$$2\angle 20° + 3\angle 40° = 2 \cos 20° + j2 \sin 20° + 3 \cos 40°$$
$$+ j3 \sin 40°$$

$$= 4.18 + j2.61$$

$$= \left| \sqrt{(4.18^2 + 2.61^2)} \right| \angle \left(\tan^{-1} \frac{2.61}{4.18} \right)$$

$$= 4.93\angle 32°$$

Example

The impedance Z in an electrical circuit is given by the following expression. Simplify the expression and obtain the impedance in the Cartesian and polar forms.

$$\frac{1}{Z} = \frac{1}{j\omega L} + j\omega C + \frac{1}{R}$$

If the $1/j\omega L$ term is multiplied by j/j then it becomes $-j/\omega L$. The expression can then be written as

$$\frac{1}{Z} = \frac{1}{R} - j\left(\frac{1}{\omega L} - \omega C \right)$$

Thus we can write

$$Z = \frac{1}{\frac{1}{R} - j\left(\frac{1}{\omega L} - \omega C \right)}$$

If we multiply the numerator and denominator by the conjugate of the denominator then we obtain

$$Z = \frac{\frac{1}{R} + j\left(\frac{1}{\omega L} - \omega C \right)}{\frac{1}{R^2} + \left(\frac{1}{\omega L} - \omega C \right)^2}$$

This is in Cartesian form, $a + jb$, and can be converted to polar form $\left| \sqrt{(a^2 + b^2)} \right| \angle \tan^{-1}(b/a)$, to give

$$Z = \frac{1}{\frac{1}{R^2} + \left(\frac{1}{\omega L} - \omega C \right)^2} \angle \tan^{-1}\left(\frac{1}{\omega L} - \omega C \right)$$

Review problems

9   If $z_1 = 3 - j2$ and $z_2 = 1 + j2$, determine:
    (a) $z_1 + z_2$, (b) $z_2 - z_1$, (c) $z_1 - z_2$, (d) $z_1 z_2$, (e) z_1/z_2,
    (f) $z_1(z_1 + z_2)$, (g) $(z_1/z_2)^2$.
10 If $z = 4\angle 20°$, determine (a) $1/z$, (b) z^2, (c) $1/z^2$.
11 Simplify, giving the answer in polar form,

$$\frac{(2\angle 10°) \times (4\angle 20°)}{5\angle 60°}$$

12 Simplify, giving the answer in Cartesian form,

$$\frac{1}{5 - j3} - \frac{1}{5 + j3}$$

Further problems

13 Determine (i) $z_1 + z_2$, (ii) $z_1 - z_2$, for the following complex
    numbers:
    (a) $z_1 = 2 + j3, z_2 = 3 + j2$,
    (b) $z_1 = 4 + j2, z_2 = 2 - j5$,
    (c) $z_1 = 2 - j3, z_2 = 2 + j5$,
    (d) $z_1 = -3 + j5, z_2 = 2 - j3$,
    (e) $z_1 = 2 - j1, z_2 = -3 + j5$,
    (f) $z_1 = -4 + j6, z_2 = -1 + j2$.
14 Determine the products of the following complex numbers:
    (a) $3 + j4$ and $5 - j2$, (b) $1 - j2$ and $4 + j2$,
    (c) $3 - j2$ and $2 - j4$, (d) $-2 + j4, 3 - j4$,
    (e) $5 + j2$ and $2 + j10$, (f) $2 - j3$ and $3 - j2$.
15 Determine the products of the following complex numbers:
    (a) $20\angle 40°$ and $3\angle 15°$, (b) $3\angle 30°$ and $4\angle 100°$,
    (c) $5\angle(-30°)$ and $4\angle 40°$, (d) $6\angle(-50°)$ and $2\angle(-60°)$,
    (e) $10\angle 10°$ and $6\angle(-10°)$, (f) $4\angle 40°$ and $5\angle 40°$.
16 Determine the results of multiplying the following complex
    numbers by their conjugates:
    (a) $5 + j1$, (b) $1 + j5$, (c) $2 - j3$, (d) $-4 - j5$.
17 Determine, in Cartesian form, the value of z_1/z_2 for the
    following:
    (a) $z_1 = 2 + j2, z_2 = 5 + j4$,
    (b) $z_1 = 1 + j2, z_2 = 4 - j1$,
    (c) $z_1 = 1 - j3, z_2 = 4 + j2$,
    (d) $z_1 = -3 + j3, z_2 = 2 - j2$.
18 Determine, in polar form, the value of z_1/z_2 for the following:
    (a) $z_1 = 1\angle 40°, z_2 = 2\angle 40°$,
    (b) $z_1 = 9\angle 10°, z_2 = 3\angle(-50°)$,
    (c) $z_1 = 2\angle(-20°), z_2 = 4\angle 60°$,
    (d) $z_1 = 2\angle 30°, z_2 = 1\angle(-40°)$.
19 If $z_1 = 1 + j3, z_2 = 2 + j4$ and $z_3 = 3 - j2$, determine:
    (a) $z_1 + z_3$, (b) $z_1 z_2$, (c) z_1/z_3, (d) $(1/z_1) + (1/z_2)$, (e) $z_1 z_2 z_3$.

20 Simplify, in Cartesian form,

$$\frac{1}{z} = \frac{1}{4+j3} + \frac{1}{1+j2}$$

21 If $z_1 = 1 + j2$, $z_2 = 2 - j3$ and $z_3 = 1 - j2$, determine:
(a) $z_1 + z_2$, (b) $z_2 - z_3$, (c) $z_1 z_2$, (d) z_1/z_3, (e) $(z_1 + z_3)z_2$,
(f) $(z_1 + z_2)/z_3$, (g) $(z_2/z_3) + z_1$.

22 If $z = 5\angle 30°$, determine (a) $1/z$, (b) z^2, (c) $1/z^2$.

23 If $z_1 = 1 + j1$ and $z_2 = 1 - j1$, determine, in both Cartesian and polar forms:
(a) $z_1 + z_2$, (b) z_1/z_2, (c) $z_1 - z_2$.

24 If $z_1 = 2 + j1$ and $z_2 = 1 - j2$, determine, in both Cartesian and polar forms:
(a) $z_1 + z_2$, (b) z_1/z_2, (c) $z_1 - z_2$.

25 Simplify, giving the result in polar form, $2\angle 30° - 4\angle 50°$.

26 Simplify, giving the results in polar form,

(a) $\dfrac{4\angle 20°}{2\angle 60°} + 3\angle 10°$, (b) $(4\angle 10°) \times (3\angle 40°) + 2\angle(-30°)$

27 Simplify, giving the results in both Cartesian and polar forms, the following expressions arising for electrical circuit analysis:

(a) $Z = \dfrac{1}{j\omega C}$, (b) $Z = R + \dfrac{1}{j\omega C}$, (c) $Z = j\omega L + \dfrac{1}{j\omega C}$,

(d) $Z = R + j\omega L + \dfrac{1}{j\omega C}$

3 Phasors

3.1 Phasors

This section is an explanation of phasors and how they simplify the analysis of circuits involving sinusoidally varying voltages and currents. The next section in this chapter is then concerned with how complex numbers can be used to represent phasors and enable algebraic manipulation of them, the remainder of the chapter being examples of how circuits can be analysed using the complex number representation of phasors.

3.1.1 Alternating voltages and currents

For a voltage v which varies sinusoidally with time we can write an equation of the form

$$v = A \sin 2\pi f t$$

where A is the amplitude of the voltage, i.e. its maximum value, f the frequency with which it alternates and t the time. To clearly indicate that it is the maximum value of the voltage, V_m, in the equation we can write it as

$$v = V_m \sin 2\pi f t$$

Such a voltage has the value 0 when the time t is 0, as in figure 3.1(a). We can imagine such a form of signal being produced by a line, of length V_m, rotating at a constant rate of revolution such that f complete revolutions are produced per second. Since there are 2π radians in one complete revolution, then the number of radians swept out per second is $2\pi f$. The line thus revolves with a constant angular velocity $\omega = 2\pi f$. Because the angular velocity is just 2π times the frequency, it is often referred to as the *angular frequency*. Hence we can write the above equation as

$$v = V_m \sin \omega t \qquad [1]$$

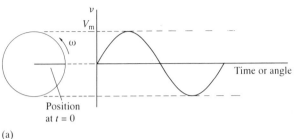

(a)

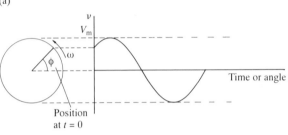

(b)

Fig. 3.1 (a) $v = V_m \sin \omega t$,
(b) $v = V_m \sin(\omega t + \phi)$

A point of detail, the policy is used of representing the instant-aneous values of quantities which vary with time by lower case letters, e.g. v, while quantities which do not vary with time are represented by capital letters, e.g. V_m.

However, we could have a voltage which varies sinusoidally with time but starts at $t = 0$ with some value, as in figure 3.1(b). For such a voltage we can write the equation describing its variation with time as

$$v = V_m \sin(2\pi ft + \phi)$$

where ϕ is what is termed the *phase difference*. This is represented by a line, of length V_m, rotating with a constant angular velocity $\omega = 2\pi f$ but starting at angle ϕ.

$$v = V_m \sin(\omega t + \phi) \tag{2}$$

Equations [1] and [2] describe how the quantity v varies with time and are said to be the *time-domain representation* of that quantity.

These lines, in figures 3.1(a) and (b), which rotate with a constant angular velocity represent by their lengths the maximum voltages, i.e. amplitudes, and their angles to the reference axis the phases of alternating voltages and currents. These lines are known as *phasors*. A phasor can be considered to be a snap-shot of the rotating line at the time $t = 0$. Such a representation of an alternating current or voltage is said to be the *frequency-domain representation*.

Phasors are drawn with their phase angles relative to some reference axis. A phasor lying on the reference axis is often termed the *reference phasor*. It is the quantity against which all the phases of the other quantities are measured. If we imagine a phasor of length 1 drawn on this axis, i.e. a unit reference phasor, then any other phasor can be considered to be just the unit reference phasor rotated through an angle and scaled in size. In series circuits it is customary to use the current as the reference phasor since the current is common to all parts of the circuit. In a parallel circuit, since the voltage is common to all parts of the circuit, the voltage is used as the reference phasor.

When we have waves of two sinusoidal quantities of the same frequency, e.g. the current and voltage in a particular circuit, then if we look at plots of them against time we find that any phase difference remains constant throughout time. The associated phasors rotate at the same angular frequency and so the angle between them remains constant.

A convenient way of specifying a phasor is by the use of polar notation. Consider a unit reference phasor. In polar notation we can represent this by $1\angle 0°$. For a phasor at an angle ϕ we just rotate the unit reference by ϕ. For a phasor of length V_m we just scale the length of the unit phasor by V_m. Thus

$$\mathbf{V} = (V_m \times 1)\angle(\phi + 0°) = (V_m \angle \phi) \times (1 \angle 0°)$$

It is the unit reference phasor we can think of as being the part that rotates, all the $V_m \angle \phi$ terms does is give the scaling factors. Generally the unit reference phasor is omitted and we just write, for a phasor of length V_m at a phase angle ϕ, measured anti-clockwise from the reference axis,

$$\mathbf{V} = V_m \angle \phi \tag{3}$$

If the phase angle was measured in a clockwise direction from the reference axis it would be a negative angle. Bold print is often used to distinguish phasor quantities from other quantities which have only the attribute of size necessary to specify them.

Although so far the length of the phasor has represented the maximum value of the quantity it is more customary to draw the length as representing the root-mean-square value, the r.m.s. value being the maximum value divided by $\sqrt{2}$. The r.m.s. value is directly proportional to the maximum value and thus the phasor drawn with the r.m.s. value is just a scaled version of the one drawn using the maximum value. All the techniques discussed for phasors with lengths representing maximum values apply equally well when the lengths are representing r.m.s. values. In this book, the maximum values of quantities are represented by always using a subscript m with the quantity as a capital letter, e.g. V_m, while r.m.s. are written as just the quantity as a capital letter with no subscript, e.g. V.

Phasors are often referred to as vectors. However, it should be pointed out that phasors and vectors differ in one important aspect. A vector diagram, e.g. that for an object subject to the action of a number of concurrent forces, shows the vectors without regard to time. They are always at those angles. A phasor diagram, however, shows the phasors at just one instant of time. They rotate with time. The types of mathematics used with vectors can however be applied to phasors. Thus, for example, the parallelogram of forces can be used to obtain the resultant force acting on an object when subject to two forces. Similarly, we can use the parallelogram method to determine the resultant phasor when, for example, we have to add two voltages to obtain the total voltage across two series components.

Example

An alternating voltage varies with time according to the equation $v = 10 \sin(100t + 30°)$ V. What phasor can be used to represent it?

The maximum amplitude of the alternating voltage is 10 V and it has a phase angle of 30°, thus the phasor, when written in terms of the amplitude is $V = 10\angle 30°$ V. However, if we express the phasor length in terms of the root-mean-square value of the voltage $V = (10/\sqrt{2})\angle 30°$ V r.m.s.

Example

What will be the time-domain representation of the voltage described by the phasor $6.0\angle 80°$ V. Assume that the length of the phasor has been expressed in terms of the r.m.s. value.

The amplitude of the alternating voltage will be $6.0 \times \sqrt{2} = 8.5$ V and the phase angle 80°. The frequency of the alternating voltage is not known, not being specified by the phasor representation. Thus the time-domain representation is $v = 8.5 \sin(\omega t + 80°)$ V.

Review problems

1　Represent each of the following with an equation in the time-domain and as a phasor with its r.m.s. value:
(a) A voltage varies sinusoidally with time, having a maximum amplitude of 4 V, a frequency of 50 Hz, and a phase angle of 60°.
(b) A current varies sinusoidally with time, having a maximum amplitude of 0.5 A, a frequency of 50 Hz, and a phase angle of 20°.
(c) A voltage varies sinusoidally with time, having a maximum amplitude of 10 V, a frequency of 1 kHz, and a phase angle of 90°.

(d) A current varies sinusoidally with time, having a r.m.s. value of 2 A, a frequency of 50 Hz, and a phase angle of 60°.

2   For each of the following, give its representation as a phasor with the r.m.s. value being used:

(a) An alternating current is represented by the equation $i = 20 \sin(200t + 40°)$ mA.

(b) An alternating voltage is represented by the equation $v = 10 \sin(377t - 30°)$ V.

(c) An alternating current is represented by the equation $i = 20.2 \sin(1000t - 120°)$ mA.

(d) An alternating voltage is represented by the equation $v = 5 \sin 500t$ V.

3   For each of the following, give the time-domain representation of the quantities described by the phasors (the phasors have been written in terms of their r.m.s. values):

(a) $3.0\angle 20°$ mA, (b) $4.0\angle(-60°)$ V, (c) $1.2\angle 90°$ A,
(d) $100\angle(-30°)$ V, (e) $50\angle 0°$ mA.

3.1.2 Adding phasors

In general, when we have a circuit which is connected to a source of a single frequency, then in the steady state all the currents and potential differences in that circuit vary sinusoidally with the same frequency. Thus, for example, if we have a voltage input at a frequency of 50 Hz then the potential differences across each of the components in the circuit and the currents will all have this

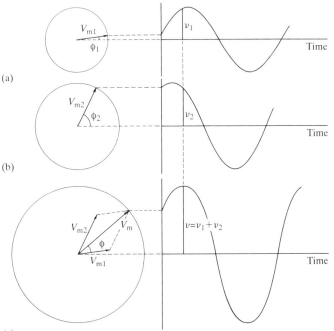

Fig. 3.2 Adding phasors

same frequency of 50 Hz. However, all the potential differences will not necessarily have the same amplitudes and the same phase.

If we want to know how the sum of the potential differences across, say, two components in series varies with time then we can add the two equations describing how their individual potential differences vary with time, say $v_1 = V_{m1} \sin(\omega t + \phi_1)$ and $v_2 = V_{m2} \sin(\omega t + \phi_2)$. Figures 3.2(a) and (b) show the graphs of the two voltages. To find the sum of the two at each instant of time, i.e. $v = v_1 + v_2$, we can add the two sine expressions or just add corresponding time ordinates on the two graphs to obtain the result shown in figure 3.2(c). However, this can be generated by a phasor which is obtained by adding the two phasors as though they were vector quantities, i.e. by using the parallelogram of vector to obtain the resultant. The resultant phasor has a length V_m and a phase ϕ, these being the length of the diagonal of the parallelogram and its angle to the reference axis. The resultant potential difference is thus

$$v = V_m \sin(\omega t + \phi)$$

The length A of this resultant phasor is given, see figure 3.3, by the use of Pythagoras theorem as

$$V_m^2 = (V_{m1} \sin\phi_1 + V_{m2} \sin\phi_2)^2 + (V_{m1} \cos\phi_1 + V_{m2} \cos\phi_2)^2$$

The phase angle ϕ of this resultant phasor is

$$\phi = \tan^{-1}\left(\frac{V_{m1} \sin\phi_1 + V_{m2} \sin\phi_2}{V_{m1} \cos\phi_1 + V_{m2} \cos\phi_2}\right)$$

Thus the sum of two, or more, phasors can be obtained by adding them as though they were vector quantities.

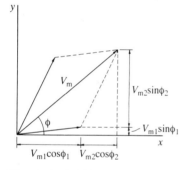

Fig. 3.3 Adding phasors

Example

The currents through two parallel arms of a circuit are given by $i = 10\sqrt{2} \sin\omega t$ mA and $i = 20\sqrt{2} \sin(\omega t + 90°)$ mA. What is the total current entering the parallel arrangement?

We can draw the currents as phasor quantities, one having a length equivalent to 10 mA and along the reference axis and the other with a length equivalent to 20 mA at 90° anticlockwise from the axis. These length values for the phasors assume that we are representing them by their r.m.s. values. Figure 3.4 shows the phasors. The resultant phasor is then obtained by treating the phasor quantities as though they were vector quantities and using the parallelogram of vectors method to determine the resultant.

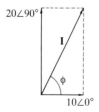

Fig. 3.4 Example

The resultant can be obtained from a scale drawing of the vectors and the parallelogram or by calculation from the parallelogram.

Using the Pythagoras theorem

$$I = \sqrt{20^2 + 10^2} = 22.4 \, \text{mA}$$

$$\phi = \tan^{-1}\left(\frac{20}{10}\right) = 63.4°$$

Thus $\mathbf{I} = 22.4\angle 63.4°$ mA and so $i = 22.4\sqrt{2} \, \sin(\omega t + 63.4°)$ mA.

Review problems

4   Determine graphically the sum of the potential differences across two series components if the potential difference across one of them is given by $2\sin\omega t$ V and across the other by $4\sin(\omega t + 90°)$ V.
5   Determine graphically the sum of the currents through two parallel arms of a circuit if the currents through each arm are given by $3\sin\omega t$ mA and $4\sin(\omega t + 90°)$ mA.

3.1.3 Subtracting phasors

Suppose we have a situation where the current entering parallel arms of a circuit is $i = I_m \sin(\omega t + \phi)$, i.e. a phasor $\mathbf{I}$. If the current through one of the arms is $i_1 = I_{m1} \sin(\omega t + \phi_1)$, i.e. a phasor $\mathbf{I}_1$, then if we need to determine the current i_2 through the other parallel arm we have to obtain $i_2 = i - i_1$. Subtracting i_1 from i is the same as adding $(-i_1)$ to i. We can therefore do this graphically by adding these currents at each instant of time or by adding the phasors $(-\mathbf{I}_1)$ and $\mathbf{I}$. A negative version of a current is one that is 180° out-of-phase, thus the phasor for the negative current is in exactly the opposite direction to that of the original phasor. Figure 3.5 shows the phasors and the resultant obtained by adding $(-\mathbf{I}_1)$ and $\mathbf{I}$.

Example

Determine the current through one of a pair of parallel arms of a circuit if the current entering the parallel arrangement is $10\sqrt{2} \, \sin\omega t$ mA and the current in the other parallel arm is $30\sqrt{2} \, \sin(\omega t + 90°)$ mA.

Figure 3.6 shows the phasors for the two currents, the phasor lengths representing the r.m.s. values of the currents. The current through one of the arms is obtained by adding the phasor for

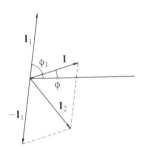

Fig. 3.5 Subtracting phasors

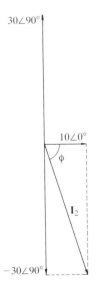

Fig. 3.6 Example

$-30\sqrt{2}\,\sin(\omega t+90°)$ mA to that for $10\sqrt{2}\,\sin\omega t$ mA. The result is

$$I_2 = \sqrt{10^2 + (-30)^2} = 31.6\,\text{mA}$$

$$\phi = \tan^{-1}\left(\frac{-30}{10}\right) = -71.6°$$

The resultant current is given by the phasor $31.6\angle(-71.6°)$ mA and so the current is $i_2 = 31.6\sqrt{2}\,\sin(\omega t - 71.6°)$ mA.

Review problems

6   Determine the current through one of a pair of parallel arms of a circuit if the current entering the circuit is $20\sin\omega t$ mA and the current in the other parallel arm is $40\sin(\omega t + 90°)$ mA.

7   The potential difference across two series components is $10\sin(\omega t + 90°)$ V. If the potential difference across one of the components is $4\sin\omega t$ V, what is the potential difference across the other component?

3.2 Complex notation for phasors

Phasors can be written in polar notation as $A\angle\theta$, such notation can however be used to represent complex numbers. Indeed the way we add and subtract phasor quantities is just the same way as we handle complex quantities (see chapter 2). Thus we can represent phasors by complex quantities. The important feature of this representation is that the operation of multiplying by $j = \sqrt{-1}$ is equivalent to a rotation anticlockwise through 90°, with multiplication by j^2 being a rotation anticlockwise through 180°, by j^3 though 270° and by j^4 through 360°. Multiplication by $-j$ is equivalent to a rotation clockwise through 90°. Thus if we take the positive real number axis as the reference axis for phasors, then a phasor drawn along that axis would be $V\angle0°$ and could, in complex notation, be represented by just the real number V. A phasor $V\angle90°$ is however at right angles to the real number axis and so in complex notation is just the imaginary number jV.

In general, a phasor can be represented in the polar and Cartesian forms of complex number. If we have the phasor in polar notion, i.e. we know the phasor length and its phase angle, and if we want to express it in Cartesian notion then we can use

$$\mathbf{V} = V\cos\theta + jV\sin\theta \qquad [4]$$

If we have the phasor in Cartesian notation, i.e. the form $a + jb$, and we want to express it in terms of polar notion,

$$\mathbf{V} = V\angle\theta \qquad\qquad [5]$$

where

$$V = \sqrt{a^2 + b^2} \qquad\qquad [6]$$

and

$$\phi = \tan^{-1}\left(\frac{b}{a}\right) \qquad\qquad [7]$$

In the above and preceding discussion of phasors it is assumed that, unless the subscript m is used, the quantities used are the r.m.s. values. For a discussion of phasors in terms of the exponential form of complex notation, see chapter 4.

Example

What is the phasor $10\angle60°$ V in the Cartesian form of complex number?

Using equation [4],

$$\mathbf{V} = 10\cos60° + j10\sin60° = 5 + j8.7\,\text{V}$$

Example

What is the phasor $4 + j2$ V in the polar form of complex number?

Using equation [6],

$$V = \sqrt{4^2 + 2^2} = 4.5\,\text{V}$$

and using equation [7],

$$\phi = \tan^{-1}\left(\frac{2}{4}\right) = 26.6°$$

Hence $\mathbf{V} = 4.5\angle26.6°$ V.

Review problems

8 What are the following phasors in polar form?
   (a) $3 + j4$, (b) $4 - j3$, (c) $-2 - j2$, (d) $-1 + j2$.
9 What are the following phasors in Cartesian form?
   (a) $4\angle30°$, (b) $5\angle120°$, (c) $2\angle(-30°)$, (d) $4\angle90°$.

3.2.1 Phasor addition and subtraction

Suppose we want to add the sinusoidal potential differences across two series components in a circuit. If the two potential differences are represented by the two phasors V_1 and V_2, then these in Cartesian form are

$$V_1 = a_1 + jb_1$$

$$V_2 = a_2 + jb_2$$

The sum of the two phasors (see chapter 2 for the addition of complex numbers) is thus

$$V_1 + V_2 = a_1 + jb_1 + a_2 + jb_2$$

$$= (a_1 + a_2) + j(b_1 + b_2) \qquad [8]$$

Thus the resultant potential difference across the two has a phasor which, in Cartesian form, has a real part which is the sum of the real parts of the two constituent phasors and an imaginary part which is the sum of the imaginary parts of the two constituent phasors.

Subtraction of one phasor from another involves just the subtraction of their complex numbers when in Cartesian form (see chapter 2). Thus

$$V_1 - V_2 = (a_1 - a_2) + j(b_1 - b_2) \qquad [9]$$

Addition and subtraction of phasors in complex notation is simplest if carried out in Cartesian notation, rather than polar notation.

Kirchhoff's laws apply to the voltages and currents in a circuit at any instant of time. Thus, the voltage law means that the sum of the voltages taken round a closed loop is zero and so, with alternating voltages, we have

$$v_1 + v_2 + v_3 + \ldots = 0 \qquad [10]$$

and hence

$$V_{m1} \sin(\omega t + \phi_1) + V_{m2} \sin(\omega t + \phi_2)$$
$$+ V_{m3} \sin(\omega t + \phi_3) + \ldots = 0$$

However, we can represent these by phasors, thus

$$V_1 + V_2 + V_3 + \ldots = 0 \qquad [11]$$

Kirchoff's voltage law can thus be expressed as: the phasor sum of all the voltages around a closed loop is zero. In a similar way, Kirchhoff's current law can be expressed as: the phasor sum of all currents at a node is zero.

Example

A circuit has three components in series. If the voltages across the components are 10 V, j6 V, and 4 + j8 V, what is the total voltage, in Cartesian notation, across the three components?

Since the components are in series, the total voltage is the sum of the three voltages. Thus

$$\mathbf{V} = 10 + j6 + 4 + j8 = 14 + j14 \text{ V}$$

Example

A circuit has two components in parallel. If the currents through the components are $4\angle 60°$ A and $2\angle 30°$ A, what is the total current, in polar notation, entering the parallel arrangement?

Since the components are in parallel, the total current is the sum of the currents through the two components. For addition it is simplest to convert the phasors into Cartesian notation. Thus

$$\mathbf{I} = (4 \cos 60° + j4 \sin 60°) + (2 \cos 30° + j2 \sin 60°)$$

$$= 2 + j3.46 + 1.73 + j1.73 = 3.73 + j5.19 \text{ A}$$

In polar notation,

$$\mathbf{I} = \sqrt{3.73^2 + 5.19^2} \angle \left(\tan^{-1} \frac{5.19}{3.73} \right) = 6.39\angle 54.3° \text{ A}$$

Example

Determine the phasor difference between the two voltages $40\angle 90°$ and $40\angle 210°$.

Putting the phasors in the Cartesian form of complex notation then we have

$$\mathbf{V} = (40 \cos 90° + j40 \sin 90°) - (40 \cos 120° + 40 \sin 120°)$$

$$= 0 + j40 - 20 - j34.6 = -20 + j5.4$$

In polar notation this becomes

$$\mathbf{V} = \sqrt{20^2 + 5.4^2} \angle \left(\tan^{-1} \frac{5.4}{-20} \right) = 20.7 \angle 164.9°$$

In order to be sure of what the angle is, it is often worthwhile to sketch the Argand diagram. In this case the phasor is drawn in the quadrant with the negative real number and the positive imaginary number.

Review problems

10 A circuit has two components in series. If the voltages across them are $4\angle 0°$ V and $3\angle 60°$, what is the total voltage, in polar notation, across the components?

11 A circuit has two components in parallel. If the currents through each component are $5\angle 0°$ mA and $10\angle 45°$ mA, what is the total current, in polar notation, emerging from the parallel arrangement?

12 A circuit has two components in series. If the total voltage across the two components is $16\angle 200°$ V and that across one of the them is $10\angle 45°$ V, determine the voltage, in polar notation, across the other component?

13 A circuit has two components in parallel. If the current entering the parallel circuit is $7\angle 45°$ A and that through one component is $4\angle 90°$ A, what is the current, in polar notation, through the other component?

14 A circuit has three components in series. If the voltages across them are $2 + j3$ V, $4 - j5$ V, and $j10$ V, what is the total voltage, in polar notation, across the components?

15 A circuit has three components in parallel. If the currents through them are $3 + j2$ A, $1 + j1$ A and $4 - j2$ A, what is the total current, in polar notation, entering the parallel circuit?

16 A circuit has two components in series. If the total voltage across the two is $12 + j3$ V and that across one component is $5 + j2$ V, what is the voltage, in polar notation, across the other component?

3.3 Impedance

The term impedance Z can be defined as the ratio of the phasor voltage across a component to the phasor current through it, i.e.

$$Z = \frac{\mathbf{V}}{\mathbf{I}} \qquad\qquad [12]$$

Since, in unit terms, it is volts/amps, it has the units of ohms.
In general, we have $\mathbf{V} = V\angle\theta$ and $\mathbf{I} = I\angle\phi$, thus

$$Z = \frac{V\angle\theta}{I\angle\phi} = \frac{V}{I}\angle(\theta - \phi)$$

Thus V/I is the magnitude of the impedance and $(\theta - \phi)$, i.e. the difference in phase angle between the voltage and current, is its phase angle. For the magnitude we could use the ratio of the r.m.s. values or the ratio of the maximum values, the result is the same.

Impedance is the ratio of two phasors and can be represented by a complex number, it is not however a phasor itself. Impedance is not a sinusoidally varying quantity. For this reason it is not, in this book, written in bold type, such type only being used for phasors. It can, however, be represented as a complex number and added, subtracted, multiplied and divided by the methods used for complex numbers (see chapter 2).

By describing phasors in terms of their unit reference phasor (see section 3.1.1) it is more apparent that impedance is not itself a phasor. Thus the voltage phasor is really $(V\angle\theta) \times (1\angle 0°)$ and the current phasor $(I\angle\phi) \times (1\angle 0°)$. Thus the impedance is

$$Z = \frac{(V\angle\theta) \times (1\angle 0°)}{(I\angle\phi) \times (1\angle 0°)}$$

The unit reference phasor $1\angle 0°$ cancels out. It was this that was rotating with angular velocity ω. All we are then left with is the ratio of two sets of scaling factors and these do not vary with time.

Example

The voltage across a component is $v = 2 \sin \omega t$ V and the current through it is $i = 0.5 \sin (\omega t - 40°)$ A. What is the impedance?

Using the maximum values of the quantities for the phasors, then

$$Z = \frac{\mathbf{V}}{\mathbf{I}} = \frac{2\angle 0°}{0.5\angle(-40°)} = 4\angle 40° \ \Omega$$

Example

A circuit element has an impedance of $12\angle 60°$ Ω, what will be the voltage across it when the current through it is $2\angle 45°$ A?

$$\mathbf{V} = \mathbf{I}Z = 2\angle 45° \times 12\angle 60° = 24\angle 105° \text{ V}$$

Example

A circuit element has an impedance of $4 + j2$ Ω, what will be the current through it when the voltage across it is $5 \sin (\omega t + 60°)$ V?

Since it is easier when dividing complex numbers to work with the polar form, the impedance is put into that form. Thus

$$Z = \sqrt{4^2 + 2^2} \angle \left(\tan^{-1} \frac{2}{4} \right)$$

and so

$$\mathbf{I} = \frac{\mathbf{V}}{Z} = \frac{5\angle 60°}{4.47\angle 26.6°} = 1.12\angle 33.4° \text{ A}$$

Review problems

17 Determine the impedance of circuits for which the following are the voltages and currents:
   (a) $v = 10 \sin 300t$ V, $i = 2 \sin (300t - 60°)$ A,
   (b) $\mathbf{V} = 12\angle 60°$ V, $\mathbf{I} = 3\angle(-30°)$ A,
   (c) $\mathbf{V} = 10 + j4$ V, $\mathbf{I} = 3 + j5$ A,
   (d) $v = 4 \sin (1000t + 20°)$ V, $i = 1 \sin (1000t + 50°)$ A.

18 Determine the voltages across the following components:
   (a) $Z = 20 + j5$ Ω, $i = 2\sqrt{2} \sin (\omega t + 60°)$ A,
   (b) $Z = 100\angle 90°$ Ω, $\mathbf{I} = 20\angle 30°$ mA,
   (c) $Z = 10 + j5$ Ω, $\mathbf{I} = 0.5\angle(-60°)$ A,
   (d) $Z = j5$ Ω, $i = 3\sqrt{2} \sin (1000t - 60°)$ A.

3.3.1 Series and parallel impedances

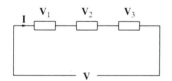

Fig. 3.7 Impedances in series

Consider the series connection of impedances, as in figure 3.7. Kirchhoff's voltage law gives

$$\mathbf{V} = \mathbf{V}_1 + \mathbf{V}_2 + \mathbf{V}_3$$

Dividing by the phasor current, which is the same for each component because they are in series, then

$$\frac{\mathbf{V}}{\mathbf{I}} = \frac{\mathbf{V}_1}{\mathbf{I}} + \frac{\mathbf{V}_2}{\mathbf{I}} + \frac{\mathbf{V}_3}{\mathbf{I}}$$

Hence the total impedance Z is

$$Z = Z_1 + Z_2 + Z_3 \qquad [13]$$

Consider the parallel connection of impedances, as in figure 3.8. Kirchhoff's current law gives

$$\mathbf{I} = \mathbf{I}_1 + \mathbf{I}_2 + \mathbf{I}_3$$

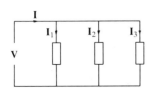

Fig. 3.8 Impedances in parallel

The voltages across each of the components is the same, thus dividing throughout by the voltage phasor gives

$$\frac{\mathbf{I}}{\mathbf{V}} = \frac{\mathbf{I_1}}{\mathbf{V}} + \frac{\mathbf{I_2}}{\mathbf{V}} + \frac{\mathbf{I_3}}{\mathbf{V}}$$

Hence the total impedance Z is given by

$$\frac{1}{Z} = \frac{1}{Z_1} + \frac{1}{Z_2} + \frac{1}{Z_3} \tag{14}$$

Example

Determine, in Cartesian form, the total impedance of a circuit consisting of impedances $2 + j3\ \Omega$, $1 + j5\ \Omega$ and $-j10\ \Omega$ in series.

Since the impedances are in series, the total impedance is

$$Z = 2 + j3 + 1 + j5 - j10 = 3 - j2\ \Omega$$

Example

Determine, in Cartesian and polar forms, the total impedance of a circuit consisting of impedances of $4\angle20°$ W and $5\angle60°$ in series.

Since the impedances are in series, the total impedance is

$$Z = 4\angle20° + 5\angle60°$$

$$= 4\cos 20° + j4\sin 20° + 5\cos 60° + j5\sin 60°$$

$$= 3.76 + j1.37 + 2.50 + j4.33$$

$$= 6.26 + j5.70$$

In polar form this is

$$Z = \sqrt{6.26^2 + 5.70^2} \ \angle\left(\tan^{-1}\frac{5.70}{6.26}\right)$$

$$= 8.47\angle42.3°\ \Omega$$

Example

Determine, in polar form, the impedance of a circuit which consists of impedances of $5\angle60°\ \Omega$ in parallel with $2\angle(-30°)\ \Omega$.

Since they are in parallel, the total impedance Z is given by

$$\frac{1}{Z} = \frac{1}{5\angle 60°} + \frac{1}{2\angle(-30°)}$$

$$= 0.2\angle(-60°) + 0.4\angle 30°$$

$$= 0.2\cos(-60°) + j0.2\sin(-60°) + 0.4\cos 30°$$
$$+ j0.4\sin 30°$$

$$= 0.10 - j0.17 + 0.35 + j0.20$$

$$= 0.45 + j0.03$$

$$= \sqrt{0.45^2 + 0.03^2}\angle\left(\tan^{-1}\frac{0.03}{0.45}\right)$$

$$= 0.45\angle 3.8°$$

Hence $Z = 1/0.45\angle 3.8° = 2.22\angle -3.8°\ \Omega$

Review problems

19 Determine, in Cartesian form, the impedances of circuits which consists of the following impedances in (i) series, (ii) parallel:
(a) 10 Ω and j10 Ω,
(b) 2 + j4 Ω and 3 + j2 Ω,
(c) $2\angle 90°\ \Omega$ and $5\angle 60°\ \Omega$,
(d) $4\angle 30°\ \Omega$ and $2\angle(-30°)\ \Omega$.

3.4 Phasors for circuit elements For a pure *resistor,* suppose the current through it is

$$i = I_m \sin\omega t$$

Since $v = Ri$, then the potential difference across it is

$$v = \frac{I_m}{R}\sin\omega t$$

or

$$v = V_m \sin\omega t$$

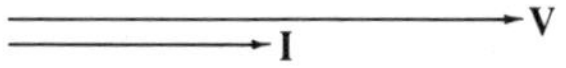

Fig. 3.9 Phasors for pure resistance

The current and voltage are thus in phase. They can be represented by phasors, see figure 3.9, expressed in terms of the maximum values of $\mathbf{I} = I_m\angle 0°$ and $\mathbf{V} = V_m\angle 0°$ or, as is more usual, in terms of the r.m.s. values as $\mathbf{I} = I\angle 0°$ and $\mathbf{V} = V\angle 0°$. The impedance Z is thus

$$Z = \frac{\mathbf{V}}{\mathbf{I}} = \frac{V\angle 0°}{I\angle 0°} = \frac{V}{I}\angle 0° = R \qquad [15]$$

For a pure resistor, since the current and voltage are in phase, the impedance is just the resistance, the real number R.

For a pure *capacitor*, consider a voltage of

$$v = V_m \sin \omega t$$

applied across it. Since $i = C\, dv/dt$, then

$$i = C\frac{d}{dt}(V_m \sin \omega t) = \omega C V_m \cos \omega t$$

Thus

$$i = I_m \sin(\omega t + 90°)$$

Fig. 3.10 Phasors for pure capacitance

The current i in the circuit leads the voltage v by 90° (or in radians $\pi/2$). The current and voltage can thus be represented by phasors, see figure 3.10, $\mathbf{I} = I\angle 90°$ and $\mathbf{V} = V\angle 0°$. The impedance Z is thus

$$Z = \frac{\mathbf{V}}{\mathbf{I}} = \frac{V\angle 0°}{I\angle 90°} = \frac{V}{I}\angle(-90°) = -j\frac{V}{I}$$

The impedance is thus just an imaginary quantity. The term *capacitive reactance* X_C is used for the ratio of the maximum, or r.m.s., voltages and current, i.e. $X_C = V_m/I_m = 1/\omega C$, and thus the above equation can be written as

$$Z = X_C\angle(-90°) = 0 - jX_C \qquad [16]$$

For a pure *inductor*, consider the current through it to be

$$i = I_m \sin \omega t$$

Since $v = L\, di/dt$, then

$$v = L\frac{d}{dt}(I_m \sin \omega t) = \omega L I_m \cos \omega t$$

Hence we can write

$$v = V_m \sin(\omega t + 90°)$$

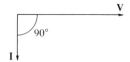

Fig. 3.11 Phasors for pure inductance

The voltage leads the current by 90° (or in radians $\pi/2$) or conversely, the current lags the voltage by 90°. The phasors representing these alternating voltages and currents (see figure 3.11) are thus $\mathbf{V} = V\angle 0°$ and $\mathbf{I} = I\angle(-90°)$. The impedance is thus

$$Z = \frac{\mathbf{V}}{\mathbf{I}} = \frac{V\angle 0°}{I\angle(-90°)} = \frac{V}{I}\angle 90° = j\frac{V}{I}$$

The impedance is thus purely imaginary. The term *inductive reactance* X_L is used for V/I and so the above equation can be written as

$$Z = X_L\angle 90° = 0 + jX_L \qquad [17]$$

The inductive reactance $X_L = V_m/I_m = \omega L$.

Notice that the real part of the impedance is resistance R and the imaginary part the reactance X. In general, if we have a component or circuit having both resistance and reactance, we can write

$$Z = R + jX \qquad [18]$$

In polar form we have

$$Z = \sqrt{R^2 + X^2}\ \angle\left(\tan^{-1}\frac{X}{R}\right) \qquad [19]$$

Thus the impedance has a magnitude of $\sqrt{R^2 + X^2}$ at an angle which has a tangent of X/R. These relationships can be represented graphically by the right-angled triangle shown in figure 3.12.

The *admittance* Y of a circuit or component is defined as the reciprocal of the impedance, i.e. $Y = 1/Z$. With impedance in ohms then admittance is in siemens (S). Admittance, in the same way as impedance, is a complex number. It can be expressed as

$$Y = G + jB \qquad [20]$$

where G is the *conductance* and B the *susceptance*. For a pure resistance

$$Y = \frac{1}{Z} = \frac{1}{R} \qquad [21]$$

There is only a real term, thus there is only conductance G of $1/R$. For a pure inductance

$$Y = \frac{1}{Z} = \frac{1}{j\omega L} = -j\frac{1}{\omega L} \qquad [22]$$

Fig. 3.12 Graphical representation of impedance

Thus there is only susceptance B of $-1/\omega L$. For a pure capacitance

$$Y = \frac{1}{Z} = j\omega C \qquad [23]$$

There is thus only susceptance B of ωC.

Example

What is the impedance, in Cartesian form, of a 2 μF capacitor when the supply voltage has a frequency of 50 Hz?

The angular frequency ω is $2\pi f$ and thus

$$Z = -j2\pi \times 50 \times 2 \times 10^{-6} = -j6.28 \times 10^{-4}\,\Omega$$

Example

What is, in polar form, the impedance of an inductor of 10 mH at an angular frequency of 1000 rad/s?

In Cartesian form

$$Z = j(1000 \times 0.010) = j10\,\Omega$$

In polar terms this becomes $Z = 10\angle 90°\,\Omega$.

Review problems

20 Determine, in Cartesian form, the impedances at a frequency of 1 kHz of the following components:
(a) 10 μF capacitor, (b) 2 H inductor, (c) 100 Ω resistor.

3.4.1 Series combinations

Consider a pure inductance in series with a pure resistance. The total impedance will be the sum of the impedances of the two elements. Thus

$$Z = R + j\omega L \qquad [24]$$

For a pure capacitance in series with a pure resistance, the total impedance will be the sum of the impedances of the two elements and thus

$$Z = R - \frac{j}{\omega C} \qquad [25]$$

or alternatively

$$Z = R + \frac{1}{j\omega C} \qquad [26]$$

For a circuit with resistance, capacitance and inductance in series then the impedance is

$$Z = R + j\omega L - \frac{j}{\omega C} = R + j\left(\omega L - \frac{1}{\omega C}\right) \qquad [27]$$

Example

What is the impedance, in Cartesian form, of the series combination of a 100 Ω resistor and a 5 mH inductor at a frequency of 100 Hz?

Since $\omega = 2\pi f$ then

$$Z = R + j\omega L = 100 + j(2\pi \times 100 \times 0.005) = 100 + j3.14 \ \Omega$$

Example

What is the impedance, in Cartesian form, of the series combination of a 1 kΩ resistor and a 10 μF capacitor at a frequency of 50 Hz?

Since $\omega = 2\pi f$ then

$$Z = R - \frac{j}{\omega C} = 1000 - \frac{j}{2\pi \times 50 \times 10 \times 10^{-6}}$$

$$= 1000 - j318 \ \Omega$$

Example

What voltage has to be supplied to a series combination of 100 Ω and a capacitive reactance of 200 Ω if a current of 0.5 A is to flow in the circuit?

Taking the current phasor as the reference phasor, then the current phasor is $0.5\angle 0°$ A. The total impedance is

$$Z = 100 - j200 \ \Omega$$

In polar form this is $\sqrt{100^2 + 200^2} \angle \tan^{-1}(-200/100)$, i.e. $223.6\angle(-63.4°)$ Ω. The voltage phasor is thus

$$\mathbf{V} = \mathbf{Z}\mathbf{I} = 223.6\angle(-63.6°) \times 0.5\angle 0° = 111.8\angle(-63.6°) \ V$$

Review problems

21 Determine, in Cartesian form, the impedances of the following series combinations of components:
   (a) 10 Ω and 1 mH at 1 kHz,
   (b) 10 Ω and 1 mH at 10 kHz,
   (c) 1 kΩ and 0.1 μF at 1 kHz,
   (d) 1 kΩ and 0.1 μF at 10 kHz.
22 What is the voltage across an inductance of 0.5 H when there is a current of $10\sqrt{2}\ \sin(100t + 30°)$ A flowing through it?
23 What is the current flowing through a capacitance of 2 μF when a voltage of $3\sqrt{2}\ \sin(60t + 45°)$ V is applied across it?
24 A circuit has a series combination of a resistance of 9 Ω, a capacitance of 1 mF and an inductance of 10 mH. What is the current when there is an input to the circuit of an alternating voltage with a phasor of $100\angle 0°$ V at an angular frequency of 100 rad/s?

3.4.2 Parallel combinations

Consider a circuit consisting of a resistance R in parallel with a capacitance C. The total impedance Z is given by

$$\frac{1}{Z} = \frac{1}{R} + \frac{1}{1/j\omega C} = \frac{1}{R} + j\omega C$$

Hence

$$Z = \frac{1}{\frac{1}{R} + j\omega C} = \frac{\frac{1}{R} - j\omega C}{\left(\frac{1}{R} + j\omega C\right)\left(\frac{1}{R} - j\omega C\right)}$$

$$Z = \frac{R}{1 + \omega^2 C^2 R^2} - j\frac{\omega C R^2}{1 + \omega^2 C^2 R^2} \qquad [28]$$

For an inductance L in parallel with a resistance R the total impedance Z is given by

$$\frac{1}{Z} = \frac{1}{R} + \frac{1}{j\omega L}$$

Hence

$$Z = \frac{1}{\frac{1}{R} + \frac{1}{j\omega L}} = \frac{\frac{1}{R} - \frac{1}{j\omega L}}{\frac{1}{R^2} + \frac{1}{\omega^2 L^2}}$$

and so

$$Z = \frac{R\omega^2 L^2}{R^2 + \omega^2 L^2} + j\frac{\omega L R^2}{R^2 + \omega^2 L^2} \qquad [29]$$

The analysis can often be made simpler with parallel circuits by working with admittance, rather than impedance. Since admittance is $1/Z$, for admittances in parallel the total admittance Y is given by

$$Y = Y_1 + Y_2 + Y_3 + \dots \qquad [30]$$

Thus for a resistance, admittance $1/R$, in parallel with a capacitance, admittance $j\omega C$,

$$Y = \frac{1}{R} + j\omega C \qquad [31]$$

For a resistance in parallel with an inductance, admittance $-j/\omega L$, then

$$Y = \frac{1}{R} - \frac{j}{\omega L} \qquad [32]$$

Example

What is the admittance of a circuit consisting of a 1 mF capacitance in parallel with a 100 Ω resistance, if the frequency is 1 kHz?

The angular frequency is $2\pi f$ and so

$$Y = \frac{1}{100} + j(2\pi \times 1000 \times 1 \times 10^{-3}) = 0.01 + j6.28 \text{ S}$$

Example

What is the impedance of a circuit with a resistance of 20 Ω in parallel with an inductance having a reactance of 10 Ω.

The inductance will have an impedance of $j10$ Ω. Thus

$$\frac{1}{Z} = \frac{1}{20} + \frac{1}{j10}$$

Hence

$$Z = \cfrac{1}{\cfrac{1}{20} + \cfrac{1}{j10}} = \cfrac{\cfrac{1}{20} - \cfrac{1}{j10}}{\cfrac{1}{20^2} + \cfrac{1}{10^2}}$$

$$= \cfrac{\cfrac{1}{20} + \cfrac{j}{10}}{0.0125} = 4 + j8 \,\Omega$$

Review problems

25 A circuit consists of an inductance with a reactance of 6 Ω in parallel with a resistance of 8 Ω, what is (a) the total impedance and (b) the current taken from the supply when a voltage of $10\sqrt{2}$ sin 100t V is supplied?

26 A circuit consists of three components in parallel, namely a resistance of 200 Ω, an inductance with a reactance of 200 Ω and a capacitance with a reactance of 100 Ω. What is the current taken from the supply when a voltage of 20$\angle$0° V is supplied?

27 A circuit consists of a resistance of 20 Ω in parallel with a capacitance of reactance 10 Ω. What is the voltage supply required if the total current to be taken from the supply is 2$\angle$0° A?

3.5 Power

When a sinusoidal voltage is applied across a pure resistance then the current is also sinusoidal, with the same frequency and in phase with the voltage. The *instantaneous power p* is thus

$$p = vi = V_m \sin \omega t \times I_m \sin \omega t$$

In general, there will be a phase difference between the current and the voltage. Thus, in general, the instantaneous power will be of the form

$$p = vi = V_m \sin \omega t \times I_m \sin (\omega t - \theta)$$

We can use the trigonometric identity

$$2 \sin A \sin B = \cos (A - B) - \cos (A + B)$$

to give

$$2 \sin \omega t \sin (\omega t - \theta) = \cos (\omega t - \omega t + \theta) - \cos (\omega t + \omega t - \theta)$$

Hence

$$p = \tfrac{1}{2}V_m I_m \cos\theta - \tfrac{1}{2}V_m I_m \cos(2\omega t - \theta)$$

The first term in the above equation does not vary with time. However, the second term describes a cosine wave with twice the frequency of the applied voltage. Since the average power of a cosine wave over a cycle is zero then the average power is just that due to the first term. Thus the *average power P* is

$$P = \tfrac{1}{2}V_m I_m \cos\theta$$

The r.m.s. voltage V is $V_m/\sqrt{2}$ and the r.m.s. current I is $I_m/\sqrt{2}$. Thus

$$P = VI \cos\theta$$

P is the termed the *real power*. The product VI is termed the *apparent power*, unit volt ampere (VA). The term $\cos\theta$ is called the *power factor*. Thus

real power = apparent power × power factor

The power factor is said to be leading when the current has a positive value of θ relative to the voltage and lagging when it is negative. We can represent these terms as the sides of a right-angled triangle, as shown in figure 3.13. The term *reactive power* Q is used for the quantity $VI \sin\theta$, the unit used being written as VAr, Var or var.

Example

A circuit consists of a resistance of 3 Ω in series with an inductive reactance of 4 Ω. Determine the apparent and real powers when a voltage of $10\angle0°$ V is applied to it. What is the power factor of the circuit?

The total circuit impedance is $3 + j4$ Ω. In polar terms this is

$$Z = \sqrt{3^2 + 4^2}\ \angle\left(\tan^{-1}\tfrac{4}{3}\right) = 5\angle53.1°\ \Omega$$

The phasor current is thus

$$\mathbf{I} = \frac{\mathbf{V}}{Z} = \frac{10\angle0°}{5\angle53.1°} = 2\angle(-53.1°)\ \text{A}$$

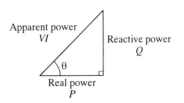

Fig. 3.13 Graphical representation of power

Thus

$$VI = 10 \times 2 = 20 \text{ VA}$$

The real power is

$$VI \cos \theta = 20 \cos (-53.1°) = 12 \text{ W}$$

The power factor is $\cos \theta$, where θ is the phase angle by which the current leads the voltage. Thus, since the current lags the voltage by 53.1°, the power factor is 0.60 lagging.

Review problems

28 A coil has an inductance of 50 mH and a resistance of 15 Ω. What will be the apparent power, the real power and the power factor when the coil is connected to a 240 V r.m.s., 50 Hz voltage supply?

29 A circuit consists of two elements in parallel, a capacitor with a reactance of 10 Ω and a resistance of 5 Ω. What will be the apparent power, the real power and the power factor when the circuit is connected to a supply of $10 \sqrt{2} \sin \omega t$ V?

3.5.1 Complex power

For power calculations, as indicated in the previous section, the real power is VI multiplied by the cosine of the phase difference between the voltage and current. However, if we multiply the phasor voltage and phasor current then we end up with

$$\mathbf{VI} = (V \angle \theta_V) \times (I \angle \theta_I) = VI \angle (\theta_V + \theta_I)$$

The product therefore has a magnitude equal to the apparent power but an angle which is the sum of the phase angles of the voltage and current. This is not what is required for power calculations. The power calculations considered above depend on the phase angle between the current and the voltage, i.e. their difference in phase angle. To obtain the difference in angles we need to work with the product of the phasor voltage and the conjugate of the current.

$$\mathbf{VI}^* = V \angle \theta_V \times I \angle (-\theta_I) = VI \angle (\theta_V - \theta_I)$$

The conjugate is indicated by the *. This product is called the *complex power* (*S*). See section 4.4.2 for a discussion of complex power in terms of the exponential forms of current and voltage phasors. The unit of the complex power is volt-amp (VA).

If we transform the above phasors into Cartesian form then, since $\cos(-A) = \cos A$ and $\sin(-A) = -\sin A$,

$$\mathbf{VI^*} = V(\cos\theta_V + j\sin\theta_V) \times I(\cos\theta_I - j\sin\theta_I)$$

$$\begin{aligned} = VI[&\cos\theta_V\cos\theta_I + \sin\theta_V\sin\theta_I) \\ &+ j(\sin\theta_V\cos\theta_I - \cos\theta_V\sin\theta_I) \end{aligned}$$

However,

$$\cos(A - B) = \cos A \cos B + \sin A \sin B$$

$$\sin(A - B) = \sin A \cos B - \cos A \sin B$$

Thus

$$\mathbf{VI^*} = VI[\cos(\theta_V - \theta_I) + j\sin(\theta_V - \theta_I)]$$

If θ is the difference in phase angle between the current and the voltage, then the complex power is

$$S = \mathbf{VI^*} = VI\cos\theta + j\,VI\sin\theta$$

Thus the real part of the complex power is the real power P and the imaginary part is the reactive power Q. The magnitude of the complex power is the apparent power.

$$S = \mathbf{VI^*} = P + jQ$$

When we have the current phasor leading the voltage phasor then Q is negative, when the current phasor lags the voltage phasor then Q is positive.

Example

A circuit has an impedance of $3 + j4$ Ω and is supplied with a voltage of $50\sqrt{2}\,\sin(1000t + 30°)$ V. What is (a) the complex power, (b) the real power, (c) the reactive power?

(a) The impedance in polar form is $\sqrt{3^2 + 4^2}\,\angle(\tan^{-1}4/3)\,\Omega$ and thus the circuit current phasor is

$$\mathbf{I} = \frac{50\angle 30°}{5\angle 53.1°} = 10\angle(-23.1°)\text{ A}$$

The conjugate of this current is $10\angle 23.1°$ A. Hence the complex power is

$$S = 50\angle 30° \times 10\angle 23.1° = 500\angle 53.1°\text{ VA}$$

In Cartesian form this is

$$S = 500 \cos 53.1° + \text{j}500 \sin 53.1° = 300 + \text{j}400 \text{ VA}$$

(b) The real power is the real part of the complex power and is thus 300 W.
(c) The reactive power is the imaginary part of the complex power and is thus 400 VAr.

Review problems

30 A voltage of $10\angle 0°$ V is applied to a circuit having an impedance of $3 + \text{j}4$ Ω. What will be the complex power, the true power and the reactive power?
31 The phasor current in a circuit is $10\angle 30°$ A when the voltage across it is $100\angle 60°$ V. Determine the complex power.

Further problems

32 State the phasors that can be used to represent the following time-domain currents and voltages. The lengths of the phasors should be expressed as r.m.s. currents and voltages.
(a) $v = 5 \sin \omega t$ V, (b) $v = 10 \sin(\omega t + 90°)$ V,
(c) $i = 20 \sin(\omega t + 60°)$ mA, (d) $i = 2 \sin(\omega t + 120°)$ A,
(e) $v = 5 \sin(\omega t - 30°)$ V, (f) $i = 20 \sin(\omega t - 90°)$ mA.
33 Write the time-domain representations of the following phasors:
(a) $20\angle 30°$ V r.m.s., (b) $150\angle 120°$ mA r.m.s.,
(c) $3\angle(-20°)$ V r.m.s., (d) $0.40\angle 0°$ A r.m.s.
34 What are the following phasors in polar form?
(a) $2 - \text{j}3$, (b) $-3 + \text{j}5$, (c) $5 - \text{j}8$, (d) $3 + \text{j}5$.
35 What are the following phasors in Cartesian form?
(a) $3\angle(-90°)$, (b) $5\angle 20°$, (c) $2\angle(-60°)$, (d) $4\angle 145°$.
36 A circuit has two components in series. If the voltages across them are $6\angle 0°$ V and $3\angle 30°$ V, what is the total voltage, in polar notation, across the components?
37 A circuit has two components in parallel. If the currents through each component are $2\angle 60°$ mA and $10\angle 0°$ mA, what is the total current, in polar notation, emerging from the parallel arrangement?
38 A circuit has two components in series. If the total voltage across the two components is $12\angle 60°$ V and that across one of them is $10\angle 30°$ V, determine the voltage, in polar notation, across the other component?
39 A circuit has two components in parallel. If the current entering the parallel circuit is $5\angle 45°$ A and that through one component is $6\angle 90°$ A, what is the current, in polar notation, through the other component?

40 A circuit has three components in series. If the voltages across them are $5 + j3$ V, $4 - j2$ V, and $j12$ V, what is the total voltage, in polar notation, across the components?

41 A circuit has three components in parallel. If the currents through them are $5 + j2$ A, $1 + j6$ A and $1 - j2$ A, what is the total current, in polar notation, entering the parallel arrangement?

42 A circuit has two components in series. If the total voltage across the two is $10 - j3$ V and that across one component is $3 + j2$ V, what is the voltage, in polar notation, across the other component?

43 A circuit has three components in parallel. The currents through two of the components are $4 + j6$ A and $2 - j2$ A. If the total current entering the parallel arrangement is $5 + j2$ A, what is the current, in Cartesian notation, through the third component?

44 A circuit has three components in series. The voltages across two of the components are $-j10$ V and 10 V. If the total voltage across the arrangement is $6 + j4$ V, what is the voltage, in Cartesian notation, across the third component?

45 Determine the impedances for the circuits for which the following are the voltages and currents:
(a) $v = 12 \sin (400t + 50°)$ V, $i = 50 \sin 400t$ mA,
(b) $v = 50 \sin 1000t$ V, $i = 5 \sin (1000t - 60°)$ A,
(c) $\mathbf{V} = 20\angle 90°$ V, $\mathbf{I} = 4\angle -30°$ A,
(d) $\mathbf{V} = 20 + j10$ V, $\mathbf{I} = 2 + j4$ A.

46 Determine the voltages across the following components:
(a) $Z = 100\angle 120°$ Ω, $i = 0.5\sqrt{2} \ \sin (500t + 40°)$ A,
(b) $Z = 40\angle(-40°)$ Ω, $\mathbf{I} = 0.2\angle 30°$ A,
(c) $Z = 20 + j5$ Ω, $i = 0.5\sqrt{2} \ \sin (1000t - 30°)$ A,
(d) $Z = 10$ Ω, $\mathbf{I} = 50\angle 120°$ mA.

47 Determine the total impedances, in Cartesian notation, of circuits which have the following impedances in series:
(a) 10 Ω and $2 + j5$ Ω,
(b) $4 + j5$ Ω and $6 - j2$ Ω,
(c) $20\angle 20°$ Ω and $10\angle 60°$ Ω,
(d) $100 - j50$ Ω and $40 + j40$ Ω.

48 Determine the total impedances, in Cartesian notation, of circuits which have the following impedances in parallel:
(a) 100 Ω and $j50$ Ω,
(b) $10 + j5$ Ω and $5 + j2$ Ω,
(c) $-j4$ Ω and 20 Ω,
(d) $j40$ Ω and $j20$ Ω.

49 A circuit consists of two impedances in series. If these impedances are $20\angle 30°$ Ω and $15\angle(-10°)$ Ω what is (a) the total impedance, (b) the circuit current, and (c) the voltage across each impedance if a voltage source of $12\angle 0°$ V is supplied to the circuit?

50 Determine the impedances of the following elements at a frequency of 1 kHz:
(a) resistance of 1 kΩ,
(b) inductance of 10 mH,
(c) capacitance of 1 mF,
(d) resistance of 1 kΩ in series with inductance of 1 mH,
(e) resistance of 1 kΩ in series with capacitance of 1 mF,
(f) series combination of a resistance of 1 kW, an inductance of 1 mH and a capacitance of 1 mF.

51   What voltage has to be supplied to a series combination of 120 Ω   resistance and a capacitive reactance of 250 Ω if a current of 0.8 A is to flow in the circuit? What is the voltage across each component?

52 What is the voltage across an inductance of 0.5 H when there is a current of $5\sqrt{2} \sin(100t - 60°)$ A flowing through it?

53 What is the current flowing through a capacitance of 2 μF when a voltage of $12\sqrt{2} \sin(60t + 20°)$ V is applied across it?

54 A circuit consists of three components in parallel, namely a resistance of 10 Ω, a capacitance of 100 μF and an inductor having a resistance of 10 Ω and an inductance of 10 mH (consider the resistance to be in series with the inductance). What voltage should be supplied to the circuit for the current taken by it to be $10\sqrt{2} \sin 1000t$ A?

55 A circuit consists of two components in parallel, a capacitance of 50 μF and an inductor having an inductance of 10 mH and a resistance of 25 Ω (consider the inductance to be in series with the resistance). What will be the current taken from the supply when a voltage of $50\sqrt{2} \sin 400t$ V is applied to the circuit?

56 For the circuit shown in figure 3.14, determine the current taken from the source.

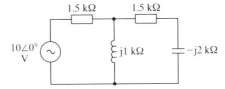

Fig. 3.14 Problem 56

57 For the circuit shown in figure 3.15, determine the current taken from the source.

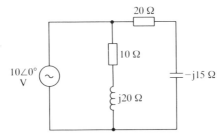

Fig. 3.15 Problem 57

58 For the circuit shown in figure 3.16, determine the current taken from the source.

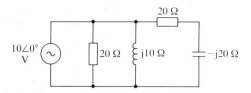

Fig. 3.16 Problem 58

59 What are the apparent power, the real power and the power factor when a circuit of impedance $15 + j2$ Ω is connected to a voltage supply of 200 V r.m.s.?

60 A circuit consists of three components in series, namely a resistance of 10 Ω, an inductance of 20 mH and a capacitance of 100 μF. What will be the apparent power, the real power and the power factor when the circuit is connected to a voltage supply of 100 sin 1000t V?

61 Determine the complex power for a circuit for which the current phasor is $2\angle 20°$ A when the voltage is $12\angle 50°$ V.

62 Determine the complex power for a circuit for which the current phasor is $0.5\angle(-40°)$ A when the voltage is $10\angle 0°$ V.

63 A resistance of 2 Ω is in series with an inductive reactance of 3 Ω. What will be the current in the circuit and the complex power developed if a voltage of $15\sqrt{2}$ sin 500t V is applied to it?

4 The exponential function

4.1 The exponential

This chapter is about the meaning that can be attached to such expressions as e^z, ln z and the trignometric functions of complex numbers like sin z and cos z.

For any real number x the exponential function e^x can be written in the form of a series as

$$e^x = 1 + x + \frac{x^2}{2!} + \frac{x^3}{3!} + \frac{x^4}{4!} + \ldots \qquad [1]$$

The series has an infinite number of terms but the terms get smaller and smaller as more terms are considered; the series is said to converge. Thus, for example, for $x = 1$ we have

$$e^1 = 1 + 1 + 0.5 + 0.166\ 67 + 0.041\ 67 + 0.008\ 33 + \ldots$$

$$= 2.716\ 67$$

If we add yet more terms then we can end up with

$$e^1 = 2.718\ 281\ 828\ 459\ 045\ 2\ldots$$

In the limit we can consider the series to give the value of e^1. Thus, in general, we can consider that in the limit the series for e^x gives the value of e^x.

Now consider this series with x replaced by a complex number z.

$$e^z = 1 + z + \frac{z^2}{2!} + \frac{z^3}{3!} + \frac{z^4}{4!} + \ldots \qquad [2]$$

Providing that the series converges then it seems reasonable to consider that in the limit it will give the value of e^z.

Example

Determine, using the series expansion for the first five terms, the value of (a) e^2, (b) e^{j1}.

(a) Using the series given in equation [1],

$$e^2 = 1 + 0.5 + \frac{0.5^2}{2!} + \frac{0.5^3}{3!} + \frac{0.5^4}{4!}$$

$$= 1 + 0.5 + 0.125 + 0.020\ 833 + 0.002\ 604 = 1.648\ 437$$

The series is converging quite rapidly so the above value is a reasonable representation of the number that would be produced by working to even more terms.

(b) Using the series given in equation [2],

$$e^{j1} = 1 + j1 + \frac{j^2 1}{2!} + \frac{j^3 1}{3!} + \frac{j^4 1}{4!}$$

Since $j^2 = -1$, $j^3 = -j$ and $j^4 = +1$, then

$$e^{j1} = 1 + j1 - 0.5 - j0.1667 + 0.0417 = 0.4583 + j0.8333$$

Review problems

1   Use the expansion as a series, for five terms, to obtain values of (a) $e^{0.1}$, (b) $e^{j0.1}$, (c) $e^{-j0.1}$.

4.1.1 The exponential form of complex numbers

A complex number z can be expressed in the form

$$z = |z|(\cos \theta + j \sin \theta)$$

Sines and cosines can also be expressed as series, with the angle θ being in radians,

$$\cos \theta = 1 - \frac{\theta^2}{2!} + \frac{\theta^4}{4!} + \ldots$$

$$\sin \theta = \theta - \frac{\theta^3}{3!} + \frac{\theta^5}{5!} + \ldots$$

Thus we can write for z,

$$z = |z| \left[\left(1 - \frac{\theta^2}{2!} + \frac{\theta^4}{4!} + \ldots \right) + j \left(\theta - \frac{\theta^3}{3!} + \frac{\theta^5}{5!} + \ldots \right) \right]$$

$$= |z|\left(1 + j\theta - \frac{\theta^2}{2!} - j\frac{\theta^3}{3!} + \frac{\theta^4}{4!} + ...\right)$$

since $j^2 = -1$, we can write the above as

$$z = |z|\left(1 + j\theta + \frac{j^2\theta^2}{2!} + \frac{j^3\theta^3}{3!} + \frac{j^4\theta^4}{4!} + ...\right)$$

However, if we write $z = j\theta$ in equation [2] then the term in brackets is $e^{j\theta}$. Thus

$$z = |z|e^{j\theta} \qquad\qquad\qquad [3]$$

This is referred to as the *exponential form* of a complex number. The polar form of a complex number $z = |z|\angle\theta$ can be regarded as a shorter way of writing the exponential form, with the angle just being referred to instead of the full exponential.

Since we have $z = |z|(\cos\theta + j\sin\theta)$ then equation [3] means that we can write

$$e^{j\theta} = \cos\theta + j\sin\theta \qquad\qquad\qquad [4]$$

This equation is often referred to as *Euler's formula*. Since

$$e^{a+jb} = e^a e^{jb}$$

then

$$e^{a+jb} = e^a(\cos b + j\sin b) \qquad\qquad\qquad [5]$$

If we have $z = |z|(\cos\theta - j\sin\theta)$ then we obtain $z = |z|e^{-j\theta}$ and so

$$e^{-j\theta} = \cos\theta - j\sin\theta \qquad\qquad\qquad [6]$$

and

$$e^{a-jb} = e^a(\cos b - j\sin b) \qquad\qquad\qquad [7]$$

Example

Determine (a) the modulus and the argument, and (b) the real and the imaginary parts, i.e. the Cartesian form, of $2\,e^{j\pi/4}$.

(a) Equation [3] gives $z = |z|e^{j\theta}$ and thus the modulus is 2 and the argument (angle) is $\pi/4$. Hence we can write $z = |2|\angle\pi/4$.

(b) Using Euler's formula (equation [4])

$$e^{j\theta} = \cos\theta + j\,\sin\theta$$

$$2\,e^{j\pi/4} = 2(\cos\pi/4 + j\,\sin\pi/4) = 1.41 + j1.41$$

Example

Determine the real and imaginary parts, i.e. the Cartesian form, of $e^{2-j\pi}$.

Using equation [7],

$$e^{a-jb} = e^{a}(\cos b - j\,\sin b)$$

$$e^{2-j\pi} = e^{2}(\cos\pi - j\,\sin\pi) = e^{2}(-1 - j0) = -e^{2}$$

There is thus only a real part.

Example

Write $4 + j4$ in the form $|z|e^{j\theta}$.

In polar form we have

$$4 + j4 = \sqrt{4^2 + 4^2}\,\angle\left(\tan^{-1}\frac{4}{4}\right) = \sqrt{32}\,\angle(\pi/4)$$

Thus the answer is $\sqrt{32}\,e^{\pi/4}$.

Review problems

2 Determine (i) the modulus and the argument and (ii) the real and imaginary parts, of (a) $5\,e^{j\pi/2}$, (b) $2\,e^{-j2\pi/3}$.
3 Determine the real and imaginary parts, i.e the Cartesian form of (a) $e^{2+j\pi/2}$, (b) $e^{4-j\pi/6}$, (c) $e^{\pi-j1}$.
4 Write the following in the form $|z|e^{j\theta}$:
(a) $j3$, (b) $2 - j2$, (c) $3 + j1$.

4.2 Trigonometric functions

Euler's formula gives us (equation [4])

$$e^{j\theta} = \cos\theta + j\,\sin\theta$$

and (equation [6])

$$e^{-j\theta} = \cos\theta - j\,\sin\theta$$

Adding these two equations gives

$$e^{j\theta} + e^{-j\theta} = 2\cos\theta$$

Thus

$$\cos\theta = \frac{e^{j\theta} + e^{-j\theta}}{2} \qquad\qquad [8]$$

Subtracting the two equations gives

$$e^{j\theta} - e^{-j\theta} = 2j\sin\theta$$

and so

$$\sin\theta = \frac{e^{j\theta} - e^{-j\theta}}{2j} \qquad\qquad [9]$$

Example

Express sin j1 as exponentials.

Using equation [9],

$$\sin\theta = \frac{e^{j\theta} - e^{-j\theta}}{2j}$$

$$\sin j1 = \frac{e^{j^2} - e^{-j^2}}{2j} = \frac{e^{-1} - e^{1}}{2j}$$

Example

Express $\cos^3\theta$ in terms of cosines of multiple angles.

Using equation [8],

$$\cos^3\theta = \left(\frac{e^{j\theta} + e^{-j\theta}}{2}\right)^3$$

$$= \tfrac{1}{8}(e^{j\theta} + e^{-j\theta})(e^{j\theta} + e^{-j\theta})(e^{j\theta} + e^{-j\theta})$$

$$= \tfrac{1}{8}(e^{j2\theta} + 2e^{0} + e^{-j2\theta})(e^{j\theta} + e^{-j\theta})$$

$$= \tfrac{1}{8}(e^{j3\theta} + 3e^{j\theta} + 3e^{-j\theta} + e^{-j3\theta})$$

$$= \frac{e^{j3\theta} + e^{-j3\theta}}{8} + \frac{e^{j\theta} + e^{-j\theta}}{8/3}$$

$$= \tfrac{1}{4}\cos 3\theta + \tfrac{3}{4}\cos\theta$$

Review problems

5   Express the following as exponentials: (a) sin 4, (b) cos j2, (c) sin j3.
6   Use the exponential forms of sines and cosines to prove that

$$\sin (A + B) = \sin A \cos B + \cos A \sin B$$

7   Express $\sin^3 \theta$ in terms of the sines of multiple angles.
8   Express $\cos^5 \theta$ in terms of the cosines of multiple angles.

4.2.1 Hyperbolic functions

The term *hyperbolic functions* is used for a family of functions involving exponentials. They are defined as follows:

$$\cosh \theta = \tfrac{1}{2}(e^\theta + e^{-\theta}) \tag{10}$$

$$\sinh \theta = \tfrac{1}{2}(e^\theta - e^{-\theta}) \tag{11}$$

$$\tanh \theta = \frac{\sinh \theta}{\cosh \theta} \tag{12}$$

Generally, cosh is pronounced as kosh, sinh as shine and tanh as than. These functions which are combinations of exponential functions have many applications in engineering. For engineers they can generally be regarded as just a shorthand way of describing combinations of exponential functions. However, it is worth noting that cos θ and sin θ can be considered to be the projections on the *x* and *y* axes of the radius of a circle when at an angle θ (figure 4.1(a)), while cosh and sinh are the projections on the *x* and *y* axes of a hyperbola (figure 4.1(b)).

Equations involving cosh, sinh and tanh can be derived to corespond with those for sines, cosines and tangents. Thus

$$\cosh \theta + \sinh \theta = \tfrac{1}{2}(e^\theta + e^{-\theta}) + \tfrac{1}{2}(e^\theta - e^{-\theta}) = e^\theta$$

$$\cosh \theta - \sinh \theta = \tfrac{1}{2}(e^\theta + e^{-\theta}) - \tfrac{1}{2}(e^\theta - e^{-\theta}) = e^{-\theta}$$

Thus

$$(\cosh \theta + \sinh \theta)(\cosh \theta - \sinh \theta) = e^0 = 1$$

and so

$$\cosh^2 \theta - \sinh^2 \theta = 1$$

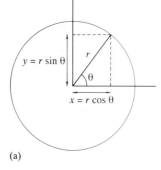

(a)

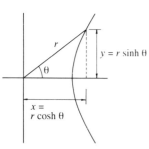

(b)

Fig. 4.1 (a) sin θ + cos θ, (b) sinh θ + cosh θ

This can be compared with the similar relationhip involving cosines and sines of

$$\cos^2\theta + \sin^2\theta = 1$$

There is a general rule, termed *Osbourne's rule* that the trigonometrical identities relating general angles involving sines, cosines and tangents may be replaced by their corresponding hyperbolic functions. However, the sign of any direct or implied product of two sines must be changed.

We can replace the term θ in the above definitions by a complex number. Then

$$\cosh j\theta = \tfrac{1}{2}(e^{j\theta} + e^{-j\theta}) = \cos\theta \qquad [13]$$

$$\sinh j\theta = \tfrac{1}{2}(e^{j\theta} - e^{-j\theta}) = j\sin\theta \qquad [14]$$

$$\tanh j\theta = \frac{\sinh j\theta}{\cosh j\theta} = j\frac{\sin\theta}{\cos\theta} = j\tan\theta \qquad [15]$$

We can also obtain, using equations [8] and [9] and replacing θ by $j\theta$,

$$\cos j\theta = \frac{e^{j^2\theta} + e^{-j^2\theta}}{2} = \frac{e^{-\theta} + e^{\theta}}{2} = \cosh\theta \qquad [16]$$

$$\sin j\theta = \frac{e^{j^2\theta} - e^{-j^2\theta}}{2j} = \frac{e^{-\theta} - e^{\theta}}{2j} = j\sinh\theta \qquad [17]$$

$$\tan j\theta = j\tanh\theta \qquad [18]$$

Example

What is the value of tanh $j\pi$?

Using equation [15],

$$\tanh j\theta = j\frac{\sin\theta}{\cos\theta} = j\frac{\sin\pi}{\cos\pi} = 0$$

Example

Show that

$$\sin(a + jb) = \sin a \cosh b + j\cos a \sinh b$$

and hence evaluate sin $(2 + j3)$.

Equation [9] gives

$$\sin\theta = \frac{e^{j\theta} - e^{-j\theta}}{2j}$$

Hence if θ is the complex number $a + jb$, then

$$\sin(a + jb) = \frac{1}{2j}(e^{j(a+jb)} - e^{-j(a+jb)})$$

$$= \frac{1}{2j}(e^{-b}e^{ja} - e^{b}e^{-ja})$$

Using Euler's formula, then

$$\sin(a + jb) = \frac{1}{2j}e^{-b}(\cos a + j\sin a) - \frac{1}{2j}e^{b}(\cos a - j\sin a)$$

$$= \sin a\left(\frac{e^{b} + e^{-b}}{2}\right) + j\cos a\left(\frac{e^{b} - e^{-b}}{2}\right)$$

$$= \sin a \cosh b + j\cos a \sinh b$$

Hence

$$\sin(2 + j3) = \sin 2 \cosh 3 + j\cos 2 \sinh 3$$

The angles are in radians, hence we have $\sin 2 = \sin 114.6° = 0.909$ and $\cos 2 = \cos 114.6° = -0.416$. For $\cosh 3$ we have $\frac{1}{2}(e^{3} + e^{-3})$ $= 10.068$ and $\sinh 3 = \frac{1}{2}(e^{3} - e^{-3}) = 10.0179$. Hence

$$\sin(2 + j3) = 0.909 \times 10.068 - j0.416 \times 10.0179$$

$$= 9.152 - j0.417$$

Review problems

9 Write (a) sinh $(-j4)$ as a sine, (b) cos $j\pi/4$ as a cosh, (c) tanh $j\pi/3$ as a tan.

10 Show that

$$\cos(a + jb) = \cos a \cosh b - j\sin a \sinh b$$

11 The input impedance Z_x of a transmission line is given by the following equation

$$Z_x = \frac{Z_0[(Z_0 + Z_R)e^{\gamma x} + (Z_R - Z_0)e^{-\gamma x}]}{(Z_0 + Z_R)e^{\gamma x} + (Z_0 - Z_R)e^{-\gamma x}}$$

Show that this can be written as

$$Z_x = \frac{Z_0(Z_0 \sinh \gamma x + Z_R \cosh \gamma x)}{Z_0 \cosh \gamma x + Z_R \sinh \gamma x}$$

4.3 The complex logarithm function

Consider the logarithm, to base e, of a complex number z. If we write

$$w = \ln z$$

then

$$z = e^w$$

We can write the complex number z as $|z|e^{j\theta}$ and w as $u + jv$. Hence

$$|z|e^{j\theta} = e^{u+jv} = e^u e^{jv}$$

Since the real parts must be equal, we have $|z| = e^u$ or $\ln |z| = u$. Since the imaginary parts must be equal, we have $e^\theta = e^v$. Hence

$$e^{v-\theta} = 1$$

However, $e^{j2n\pi} = \cos 2n\pi + j \sin 2n\pi = 1$, where n is an integer. Hence

$$v - \theta = 2n\pi$$

and so $v = \theta + 2n\pi$, where n is an integer. We thus have

$$w = u + jv = \ln |z| + j(\theta + 2n\pi)$$

Thus

$$\ln z = \ln |z| + j(\theta + 2n\pi) \qquad [19]$$

or

$$\ln (|z|e^{j\theta}) = \ln |z| + j(\theta + 2n\pi) \qquad [20]$$

The complex logarithm of a number is thus multi-valued, since there are values for each value of n.

Example

Express $\ln (2 + j1)$ in the form $a + jb$.

Expressing the complex number $2 + j1$ in exponential form gives a modulus of $\sqrt{2^2 + 1^2}$ and an argument θ which is given by $\tan^{-1} 1/2$. Hence $\theta = 26.6° = 0.46$ rad. Thus the exponential form is $\sqrt{5}\, e^{j0.46}$. Hence, using equation [20],

$$\ln(2 + j1) = \ln\sqrt{5} + j(0.46 + 2n\pi) = 2.24 + j(0.46 + 2n\pi)$$

Review problems

12 Express the following in the form $a + jb$:
  (a) $\ln(-1)$, (b) $\ln(1 - j1)$, (c) $\ln(1 - j2)$.

4.4 The phasor

A sinusoidal voltage, or current, can be expressed in the form

$$v = V_m \sin(\omega t + \phi)$$

or

$$v = V_m \cos(\omega t + \phi)$$

Euler's formula (equation [4]), however, gives

$$e^{j\theta} = \cos\theta + j\sin\theta$$

and so

$$e^{j(\omega t + \phi)} = \cos(\omega t + \phi) + j\sin(\omega t + \phi)$$

$$V_m\, e^{j(\omega t + \phi)} = V_m \cos(\omega t + \phi) + j V_m \sin(\omega t + \phi)$$

Thus we can consider the sinusoidally varying voltage to be either the real or the imaginary part of $V_m\, e^{j(\omega t + \phi)}$, depending on whether we are considering the cosine or sine function. So we can write

$$v = \mathrm{Re}\,(V_m\, e^{j(\omega t + \phi)})$$

Re is used to indicate that we are considering the real part of that which follows the symbol. The equation can be rewritten as

$$v = \mathrm{Re}\,(V_m\, e^{j\omega t} e^{j\phi})$$

The *phasor* is $V_m\, e^{j\omega t} e^{j\phi}$ and the real part of it is its projection on the real number axis.

$$\mathbf{V} = V_m\, e^{j\omega t} e^{j\phi} \qquad\qquad [21]$$

Due to the $e^{j\omega t}$ term we are said, when using phasors, to be operating in the *frequency-domain*.

When we are considering the analysis of a particular circuit, perhaps using Kirchhoff's voltage law and adding the voltages round a circuit loop, the frequency is constant. Thus the exponential term involving the angular frequency will cancel out. Consequently it is generally adequate to just work with $V_m e^{j\phi}$. Hence we can write

$$\mathbf{V} = V_m e^{j\phi} \qquad\qquad [22]$$

This leads to an abbreviated form of specification, called the polar form, in which reference is just made to the magnitude of the voltage and its phase angle, i.e. $V_m \angle \phi$. This is generally further altered by referring to the r.m.s. value of the voltage instead of the maximum value. See chapter 3 for a more detailled discussion of phasors and their use in circuit analysis.

Example

Express the voltage $v = 10 \cos(500t + 30°)$ in complex exponential form.

Using Euler's formula, as above, then

$$10e^{j(500t + 30°)} = 10\cos(500t + 30°) + j10\sin(500t + 30°)$$

Thus

$$v = \text{Real part of } 10e^{j(500t + 30°)}$$

Review problems

13 Express the following in complex exponential form:
   (a) $i = 2 \sin(100t + 45°)$,
   (b) $v = 12 \cos(500t - 60°)$,
   (c) $v = 5 \sin 300t$.

4.4.1 Impedance

The fact that impedance is not a phasor is readily apparent if the exponential forms of phasors are considered. Thus

$$Z = \frac{\mathbf{V}}{\mathbf{I}} = \frac{V_m e^{j\omega t} e^{j\phi}}{I_m e^{j\omega t} e^{j\theta}}$$

The $e^{j\omega t}$ terms cancel out and so the impedance is independent of time. Hence it is not a phasor.

4.4.2 Complex power

Complex power S is the product of the voltage phasor and the conjugate of the current phasor, i.e.

$$S = \mathbf{VI}^*$$

As equation [4] indicates,

$$e^{j\theta} = \cos\theta + j\sin\theta$$

The conjugate of this complex number is $\cos\theta - j\sin\theta$. However, equation [6] gives

$$e^{-j\theta} = \cos\theta - j\sin\theta$$

Thus the conjugate of a phasor $I_m e^{j(\omega t + \theta)}$ is $I_m e^{-(j\omega t + \theta)}$. Thus for the complex power we can write

$$S = V_m e^{j(\omega t + \phi)} I_m e^{-j(\omega t + \theta)}$$

The $e^{j\omega t}$ and $e^{-j\omega t}$ terms cancel and so we are left with

$$S = V_m I_m e^{j(\phi - \theta)}$$

or, in polar notation, $V_m I_m \angle(\phi - \theta)$, as deduced in section 3.5.1. Thus the complex power does not include a $j\omega t$ element and so is not a phasor.

Further problems

14 Determine (i) the modulus and the argument, (ii) the real and imaginary parts, of:
(a) $4\,e^{j\pi/4}$, (b) $2\,e^{j\pi/3}$, (c) $3\,e^{-j\pi/6}$.

15 Determine the real and the imaginary parts, i.e. the Cartesian form, of:
(a) $e^{j\pi}$, (b) $e^{2+j\pi/2}$, (c) $e^{4-j\pi/6}$, (d) e^{2-j2}.

16 Write the following in the form $z = |z|e^{j\theta}$:
(a) $4 + j4$, (b) $j5$, (c) $2 + j5$, (d) 4.

17 Use the exponential forms of sines and cosines to prove that

$$\cos(A + B) = \cos A \cos B - \sin A \sin B$$

18 Express $\sin^6\theta$ in terms of the cosines of multiple angles.

19 Express $\sin^7\theta$ in terms of the sines of multiple angles.

20 Write (a) $\sin j\pi/4$ as a sinh, (b) $\sinh j\pi/4$ as a sine, (c) $\cos j\pi/3$ as a cosh, (d) $\tan j\pi/3$ as a tanh.

21 Evaluate $\sin(1 + j2)$.

22 Write sinh $(3 + j4)$ in the form $a + jb$.

23 Express the following in the form $a + jb$:
(a) ln j2, (b) ln (-9), (c) ln $(2 + j4)$.

24 Express the following phasors in complex exponential form:
(a) $1 + j2$, (b) $4 - j2$, (c) $25\angle(-30°)$, (d) $10\angle40°$.

25 In an analysis of the attenuation of a T-section filter the propagation coefficient γ is $\alpha + j\beta$, where α is the attenuation coefficient and β the the phase change coefficient.
(a) Show that

$$\cosh\gamma = \cosh\alpha\cos\beta + j\sinh\alpha\sin\beta$$

(b) For a low pass filter when the frequency is less that the critical frequency, the attenuation coefficient is zero. Hence show that

$$\cosh\gamma = \cos\beta$$

(c) For the low pass filter in the attenuation band, i.e. when the frequency is greater than the critical frequency, then the phase change coefficient is π. Hence show that

$$\cosh\gamma = -\cosh\alpha$$

26 What is the attenuation α of a low-pass T-section filter at a frequency f of twice the cut-off frequency f_c if

$$\cosh\alpha = \frac{2f^2}{f_c^2} - 1$$

5 Powers of complex numbers

5.1 De Moivre's theorem

This chapter is about the handling of powers of complex numbers. Multiplication of a complex number in polar form (see section 2.3) involves the multiplication of the modulus of each of the numbers and the adding of their arguments, i.e.

$$z_1 z_2 = |z_1|\angle\theta_1 \times |z_2|\angle\theta_2 = |z_1||z_2|\angle(\theta_1 + \theta_2)$$

Thus if we have z^2, i.e. $z \times z$, then

$$z^2 = |z^2|\angle 2\theta$$

If we have z^3, i.e. $z \times z \times z$, then

$$z^3 = |z^3|\angle 3\theta$$

Hence, if we have z^n then

$$z^n = |z^n|\angle n\theta \tag{1}$$

This is known as *de Moivre's theorem*. We can also express this equation in the form

$$z^n = |z|(\cos n\theta + j \sin n\theta) \tag{2}$$

n can be either positive or negative, integer or fractional.

Example

Evaluate z^3 if $z = 1 - j1$.

In polar form

$$1-j1 = \sqrt{1^2 + 1^2} \ \angle(\tan^{-1} - 1/1) = \sqrt{2} \ \angle(-45°)$$

Hence, applying de Moivre's theorem (equation [1]),

$$z^3 = \left|(\sqrt{2})^3\right| \angle\{3 \times (-45°)\}$$

and so

$$z^3 = 2\sqrt{2}\,\{\cos(-135°) + j\,\sin(-135°)\}$$

$$= -2 - j2$$

Example

Solve the equation $z^5 = 1 + j1$.

In polar form

$$1 + j1 = \sqrt{1^2 + 1^2}\,\angle(\tan^{-1} 1/1) = \sqrt{2}\,\angle 45°$$

We need however to modify the above answer. This is because we can obtain the above complex number by a number of arguments, namely 45°, 360° + 45°, 2 × 360° + 45°, etc. Thus we should write

$$1 + j1 = \sqrt{2}\,\angle(360n + 45)°$$

where n can have the value 0, 1, 2, 3, etc.

What is required is z, i.e. $(1 + j1)^{1/5}$. Hence, applying de Moivre's theorem (equation [1]),

$$(1 + j1)^{1/5} = \left|\left(\sqrt{2}\right)^{1/5}\right| \angle\left(\tfrac{1}{5}(360n + 45)°\right)$$

Thus, with

$n = 0$ we have $1.072\angle 9°$,
$n = 1$ we have $1.072\angle 81°$,
$n = 2$ we have $1.072\angle 153°$,
$n = 3$ we have $1.072\angle 225°$,
$n = 4$ we have $1.072\angle 297°$.

With higher values of n we end up with the same values again (e.g. $n = 5$ gives $1.072\angle 369°$ which is the same as $1.072\angle 9°$). We thus have five roots for the equation. We can represent these on an Argand diagram as radial lines of length 1.072 at angles of 9°, 81°, 153°, 225° and 297°, as in figure 5.1. Note that once we have found the first root, the others are distributed round the diagram at regular intervals of $360°/n$.

In Cartesian form these roots are $1.059 + j0.168$, $0.168 + j1.059$, $-0.955 + j0.487$, $-0.758 - j0.758$, and $0.487 - j0.955$.

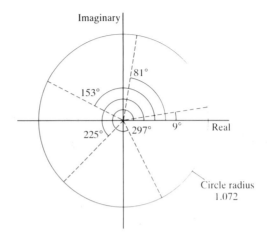

Fig. 5.1 The five roots of $(1 + j1)^{1/5}$

Example

Determine the cube roots of 5.

Putting $z = 5$, then in polar form the complex number becomes $5\angle(360n + 0)°$. Using de Moivre's theorem (equation [1]) then

$$z^{1/3} = 5^{1/3}\angle(360n/3)°$$

Thus the values are, for $n = 0$ of $1.17\angle0°$, for $n = 1$ of $1.71\angle120°$, for $n = 2$ of $1.17\angle240°$. Figure 5.2 shows the Argand diagram. In Cartesian form these are 1.17, $-0.59 + j1.01$, $-0.59 - j1.01$.

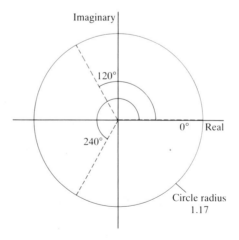

Fig. 5.2 Cube roots of 5

Example

Determine the cube root of $27\angle30°$.

Let $z = 27\angle30°$. Using de Moivre's theorem (equation [1]),

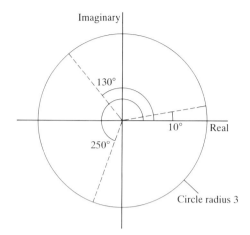

Fig. 5.3 Cube roots of $27\angle 30°$

$$z = 27^{1/3}\angle(360°n + 30°)/3$$

Thus the values are, for $n = 0$ of $3\angle 10°$, for $n = 1$ of $3\angle 130°$, for $n = 2$ of $3\angle 250°$. Figure 5.3 shows the Argand diagram.

Review problems

1 Solve the following equations, giving the answers in both polar and Cartesian forms:
   (a) $z^6 = 1$, (b) $2z^2 = -1 + j1$, (c) $z^3 = j1$, (d) $z^3 = 1 + j1$.
2 Determine the values of :
   (a) $(1 + j1)^3$, (b) $(2 + j1)^6$, (c) $(3 + j4)^3$, (d) $(-3 + j4)^3$.
3 Determine the fourth root of 16.
4 Determine the fifth root of 1.
5 Solve the following equations:
   (a) $z = (2 - j1)^{-2/3}$, (b) $z = (-2 + j3)^{-2/3}$, (c) $z = (-2 + j2)^{1/3}$.
6 Determine in polar form (a) the cube roots of $8\angle 120°$, (b) the fourth roots of $20\angle 40°$, (c) the fifth roots of $12\angle 300°$.

5.1.1 Deducing the theorem from the exponential form

De Moivre's theorem can be deduced from the exponential form of complex numbers. Thus if we have $z = |z|e^{j\theta}$, then

$$z^n = (|z|e^{j\theta})^n = |z^n|(e^{j\theta})^n = |z^n|e^{jn\theta}$$

Thus, using Euler's formula (equation [4], chapter 4),

$$z^n = |z^n|(\cos n\theta + \sin n\theta)$$

This is de Moivre's theorem.

5.2 Cosines and sines of $n\theta$

De Moivre's theorem can be used to expand $\cos n\theta$ and $\sin n\theta$, when n is some positive integer. Thus the theorem gives (equation [2])

$$z^n = |z|(\cos n\theta + j \sin n\theta)$$

However, using Euler's formula,

$$z^n = [|z|(\cos \theta + j \sin \theta)]^n$$

Thus

$$\cos n\theta + j \sin n\theta = (\cos \theta + j \sin \theta)^n \qquad [3]$$

This right-hand side of the equation can be expanded by means of the binomial series, this being of the form

$$(a+b)^n = a^n + na^{n-1}b + \frac{n(n-1)}{2!}a^{n-2}b^2 + \ldots$$

Thus

$$\cos n\theta + j\sin n\theta = \cos^n\theta + n\cos^{n-1}\theta \times j\sin\theta$$

$$+ \frac{n(n-1)}{2!}\cos^{n-2}\theta \times j^2\sin^2\theta$$

$$+ \frac{n(n-1)(n-2)}{3!}\cos^{n-3}\theta \times j^3\sin^3\theta + \ldots$$

However, $j^2 = -1$, $j^3 = -j$, etc. Thus, equating real terms gives

$$\cos n\theta = \cos^n\theta - \frac{n(n-1)}{2!}\cos^{n-2}\theta \sin^2\theta$$

$$+ \frac{n(n-1)(n-2)(n-3)}{4!}\cos^{n-4}\theta \sin^4\theta + \ldots \qquad [4]$$

and equating imaginary terms gives

$$\sin n\theta = n\cos^{n-1}\theta \sin\theta - \frac{n(n-1)(n-2)}{3!}\cos^{n-3}\theta \sin^3\theta$$

$$+ \frac{n(n-1)(n-2)(n-3)(n-4)}{5!}\cos^{n-5}\theta \sin^5\theta + \ldots \quad [5]$$

Example
Express $\cos 3\theta$ in terms of powers of cosines and sines of θ.

Using equation [4] with $n = 3$,

$$\cos 3\theta = \cos^3\theta - \frac{3(3-1)}{2!}\cos\theta\sin^2\theta$$

The next, and following, terms in the series involve terms being multiplied by $(n - 3)$ and so with $n = 3$ are zero. Thus

$$\cos 3\theta = \cos^3\theta - 3\cos\theta\sin^2\theta$$

Example

Express $\sin 3\theta$ in terms of powers of sines and cosines of θ.

Using equation [5] with $n = 3$,

$$\sin 3\theta = 3\cos^2\theta\sin\theta - \frac{3\times2\times1}{3!}\cos^{3-3}\theta\sin^3\theta$$

$\cos^0\theta = 1$ (if you are not sure of this, consider the series for the cosine, see section 4.1.1) and further terms in the series involve terms being multiplied by $(n - 3)$ and so, since $n = 3$, are zero. Thus

$$\sin 3\theta = 3\cos^2\theta\sin\theta - \sin^3\theta$$

Review problems

7   Express $\cos 4\theta$ in terms of powers of sines and cosines of θ.
8   Express $\sin 5\theta$ in terms of powers of sines and cosines of θ.
9   Express $\cos 8\theta$ in terms of powers of sines and cosines of θ.

Further problems

10 Solve the following equations:
    (a) $z^3 = 8$, (b) $z^4 = 1 - j1$, (c) $z^3 = 5 - j3$, (d) $z^2 = j1$.
11 Determine the values of:
    (a) $(1 - j\sqrt{3})^{10}$, (b) $(2 + j1)^3$, (c) $(3 - j4)^5$, (d) $(2 + j2)^4$.
12 Determine the fourth roots of (-16), i.e. $(-16)^{1/4}$.
13 Determine the square roots of $j1$.
14 Determine in polar form:
    (a) the square roots of $9\angle60°$,
    (b) the cube roots of $125\angle90°$,
    (c) the fourth roots of $81\angle80°$,
    (d) the tenth roots of $100\angle150°$.
15 Express $\sin 4\theta$ in terms of powers of sines and cosines of θ.
16 Express $\cos 6\theta$ in terms of powers of sines and cosines of θ.
17 Express $\sin 7\theta/\sin\theta$ in terms of powers of sines of θ.
18 Express $\tan 5\theta$, i.e. $\sin 5\theta/\cos 5\theta$, in terms of powers of tangents of θ.

6 Loci on the complex plane

6.1 The complex plane

The Argand diagram enables complex numbers to be represented by points on a plane, often referred to as the *complex plane*. Thus a point z can be described by either its co-ordinates x and y when it is in the Cartesian form $x + jy$ or its modulus $|z|$ and argument θ when in the polar form $|z|\angle\theta$. Figure 6.1 illustrates these two ways of describing the point.

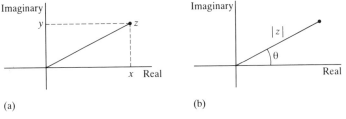

(a)   (b)

Fig. 6.1 Describing a point on the complex plane, (a) by $z = x + jy$, (b) by $z = |z|\angle\theta$

Now suppose we want to describe a set of points. We could do this by listing the values of each of the points in the set or by describing the conditions that have to be met by the z values of the points in the set. Thus, we might state that all the points must have a particular relationship between their x and y values, e.g. $x = y$, or that all the points should have a particular argument and/or modulus. A set of points which satisfy given conditions is called a *locus* (plural *loci*).

Note that in drawing a locus, the convention is generally adopted of using a solid line for a boundary which is included in the locus and a dotted line for one which is not included.

6.1.1 Lines on the complex plane

Suppose we want to describe all the points which have an argument of 45°. All these complex numbers must lie on a line which is at an angle of 45° with the real axis. Figure 6.2 shows such a line. Thus, specifying a condition of a constant argument

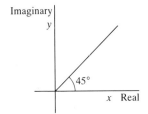

Fig. 6.2 Locus of points satisfying $\arg z = 45°$

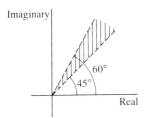

Fig. 6.3 Locus of points satisfing $45° < \arg z < 60°$

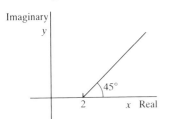

Fig. 6.4 Locus of points satisfying $\arg (z - 2) = 45°$

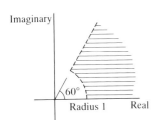

Fig. 6.5 Locus of points satisfying $0° < \arg (z - 1) < 60°$

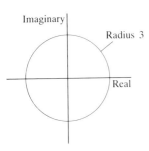

Fig. 6.6 Locus of points satisyfing $|z| = 3$

results in the specification of a series of points which lie along a line. In terms of Cartesian co-ordinates with $z = x + jy$, the condition $\arg z = 45°$ means $\tan^{-1} (y/x) = 1$ and so $y = x$. Thus another way of specifying this argument condition is as $y = x$.

If we specified the condition for the set of points of $45° < \arg z < 60°$, then all the points must lie between lines at $45°$ and $60°$. Figure 6.3 shows the situation. This condition thus specifies all the points lying within an arc.

The discussions so far have involved lines passing through the origin. Consider the condition $\arg (z - 2) = 45°$. The condition $\arg z = 45°$ gives a line at $45°$ which passes through the origin (as in figure 6.2) and is described by the equation $y = x$, with x being the real part of Z and y the imaginary part. For the condition of $\arg (z - 2) = 45°$ if we replaced the term in brackets by some other complex number w then we would have $\arg w = 45°$ and so a line at $45°$ passing through the origin. But $w = z - 2$ and so we have a line at $45°$ which starts at $(2, 0)$. It has the equation $y = x + 2$. Figure 6.4 shows the line. Thus a line which does not pass through the origin will be described by a condition of the form

$$\arg (z - c) = \theta$$

Example

Sketch the locus of points satisfying the condition:

$$0° < \arg (z - 1) < 60°$$

The condition $0° < \arg z < 60°$ describes all the points in an arc bounded by lines at $0°$ and $60°$ while that of $0° < \arg (z - 1) < 60°$ describes all the points in this arc which lie further out from the origin than $|z| = 1$. Figure 6.5 shows the resulting locus.

Review problems

1 Sketch the loci of points satisfying each of the following conditions:
(a) $30° < \arg z < 120°$,
(b) $\arg (z - 3) = 60°$,
(c) $\arg (z - 2 - j2) = 45°$,
(d) $45° < \arg (z - 2) < 90°$.

6.1.2 Circles on the complex plane

Consider a set of points described by the condition $|z| = 3$. All these complex numbers must lie on a circle of radius 3 and having its centre at the origin of the complex plane. Figure 6.6 shows such a circle. Thus specifying a condition of just a modulus results

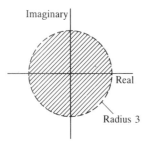

Fig. 6.7 Locus of points satisfying $|z| < 3$

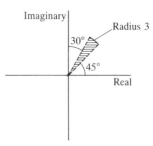

Fig. 6.8 Locus of points satisfying $30° <$ arg $z < 45°$ and $|z| \leq 3$

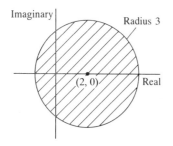

Fig. 6.9 Locus of points satisfying $|z - 2| \leq 3$

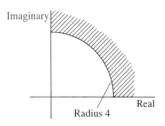

Fig. 6.10 Locus of points satisfying $0° <$ arg $z < 90°$ and $|z| \geq 4$

in the specification of a series of points which lie on a circle with its centre at the origin of the complex plane. If you are familiar with the equation in Cartesian co-ordinates of a circle, then the condition $|z| = 3$ when written in Cartesian co-ordinates is $\sqrt{x^2 + y^2} = 3$ and so $x^2 + y^2 = 3^2$. The general equation for a circle with a radius r centred on the origin is $x^2 + y^2 = r^2$. Thus the equation indicates a circle of radius 3 centred on the origin.

We could describe a set of points by the condition $|z| < 3$. This would describe all the points which lie within the circle in figure 6.7. Alternatively, the condition $|z| > 3$ describes all the points outside the circle.

Consider the locus described by the conditions that we have $45° <$ arg $z < 60°$ together with $|z| \leq 3$. The argument condition specifies all points lying between lines at $45°$ and $60°$ while the modulus condition specified all points lying inside or on a circle of radius 3 centred at the origin. The two conditions combined thus specify all points lying within the area shown in figure 6.8.

Consider the condition $|z - 2| \leq 3$. The condition $z \leq 3$ describes all the points lying within a circle of radius 3 which is centred on the origin. Since $z = x + jy$ then we can write for $|z - 2|$ the magnitude $|x + jy + 2|$ or $|(x + 2) + jy|$. This means that all the real values of z have 2 added to them and all the imaginary values have 0 added to them. As a consequence the centre of the circle is displaced from the origin to the point $(2, 0)$. Figure 6.9 shows the resulting locus. We can show this to be the case algebraically. The Cartesian form of the condition is

$$\sqrt{(x + 2)^2 + y^2} \leq 3$$

Thus we have $(x + 2)^2 + y^2 \leq 3^2$. This is the equation for a circle of radius 3 with its centre at $(2, 0)$.

Example

Sketch the locus of the points satisfying the conditions:

$$0° < \text{arg } z < 90° \text{ and } |z| \geq 4$$

The argument condition specifies all points lying between lines at $0°$ and $90°$, i.e. all the points lying within the first quadrant of the complex plane. The modulus condition specifies all points lying on or outside a circle of radius 4. Thus the locus is as shown in figure 6.10.

Example

Sketch the locus of points satisfying the conditions:

$$|z - 4 - j1| < 2 \text{ and } 0° < \text{arg } z < 90°$$

Fig. 6.11 Locus of points satisfying $|z - 4 - j1| < z$ and $0° < \arg z < 90°$

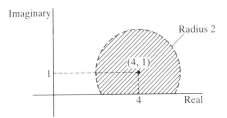

Rewriting the condition $|z - 4 - j1|$ with z as $x + jy$ results in $|x + jy - 4 - j1|$ and hence $|(x - 4) + j(y - 1)|$. Thus all the points have their x values reduced by 4 and their y values reduced by 1. As a consequence this condition describes all the points lying within a circle of radius 2 centred on the point (4, 1). The other condition $0° < \arg z < 90°$ indicates all the points lying within the arc bounded by lines at $0°$ and $90°$. Thus the locus is as shown in figure 6.11.

Review problems

2 Sketch the loci of points satisfying the following conditions:
(a) $|z| = 5$,
(b) $90° < \arg z < 180°$ and $|z| \le 2$,
(c) $0° < \arg z < 60°$ and $|z| \ge 3$,
(d) $|z + 2| = 3$,
(e) $|z - 1 + j1| = 2$,
(f) $|(z + 2 - j1)| < 2$.

6.2 Impedance diagrams

When an impedance is expressed as a complex number it can be represented on an Argand diagram by a position related to its magnitude and argument if in polar form, or its real and imaginary parts if in Cartesian form. If one of the variables determining the impedance changes then a series of points, i.e. a locus, will be produced.

Consider a circuit consisting of a resistance R in series with an inductance L when the voltage, or current, has an angular frequency ω. The impedances of the components will be as shown in figure 6.12. Hence the total impedance Z is

$$Z = R + j\omega L$$

For particular values of R, ω and L there will be a single point on an Argand diagram. However, suppose we want to consider how the impedance will change for angular frequencies varying from 0 to infinity. By inspection of the above equation we could recognise that the real part will remain unchanged and the imaginary part will

Fig. 6.12 Series *RL* circuit

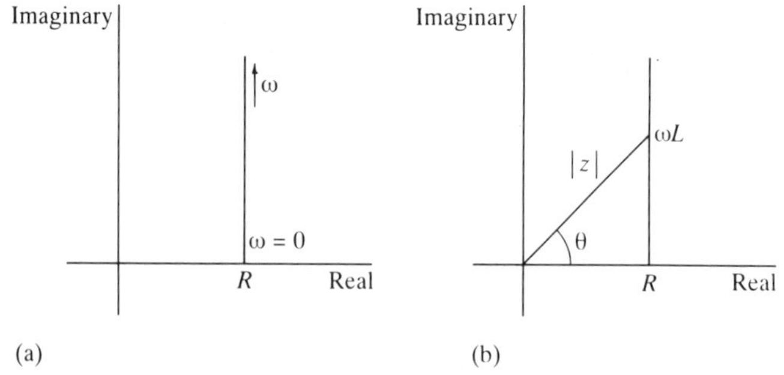

Fig. 6.13 Locus for series *RL* circuit with ω changing

(a)   (b)

increase in direct proportion to ω. Thus the locus will be as shown in figure 6.13(a). However, we can consider it in the manner described in section 6.1. Rearranging the equation, we can write it with the variable quantity isolated on the right hand side, i.e.

$$Z - R = j\omega L$$

Then

$$\arg(Z - R) = \tan^{-1}(L/0) = 90°$$

This then describes the line shown in figure 6.13(a).

If we had, say, an angular frequency of 0 then the impedance is purely R. If the angular frequency is ω then the impedance is represented by the line shown in figure 6.13(b). This has a magnitude, given by Pythagoras theorem, of $\sqrt{R^2 + (\omega L)^2}$ and an argument of $\theta = \tan^{-1}(\omega L/R)$.

Now suppose we consider the impedance of the series arrangement of R and L with ω constant but R varying from 0 to infinity. The imaginary part of the impedance will remain unchanged but the real part will vary in a linear manner. Rearranging the equation to isolate the variable on the right hand side gives

$$Z - j\omega L = R$$

Then

$$\arg(Z - j\omega L) = \tan^{-1}(0/R) = 0°$$

Thus the locus will be as shown in figure 6.14.

Fig. 6.14 Locus for series *RL* circuit with *R* changing

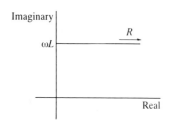

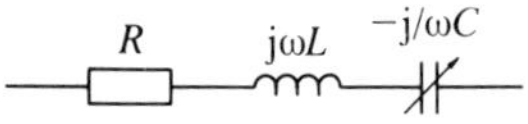

Fig. 6.15 Series *RLC* circuit

Example

A circuit consists of three components in series, a resistance R, an inductance L and a capacitance C, as in figure 6.15. Determine the locus of the points on the Argand diagram which describe how the impedance varies as the capacitance is varied.

The impedance Z of the circuit is given by

$$Z = R + j\omega L - \frac{j}{\omega C}$$

This can be written with just the variable on the right hand side of the equation as

$$Z - R - j\omega L = -\frac{j}{\omega C}$$

Thus

$$\arg(Z - R - j\omega L) = \tan^{-1}(-\frac{1/\omega C}{0}) = -90°$$

Thus the locus is as shown in figure 6.16. Note that the locus diagram clearly indicates that the argument will be zero when the impedance is equal to just R, this being the value of the intercept of the locus with the real axis.

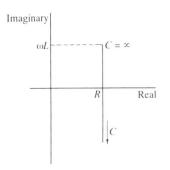

Fig. 6.16 Locus for series RLC circuit with C changing

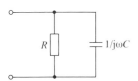

Fig. 6.17 Parallel RC circuit

Example

A circuit consists of a resistance in parallel with a capacitance, with the impedances of the components being as shown in figure 6.17. Sketch the locus of points describing how the impedance of the circuit varies with angular frequency.

The impedance Z of the circuit is given by

$$\frac{1}{Z} = \frac{1}{R} + \frac{1}{1/j\omega C}$$

Thus

$$Z = \frac{1}{\frac{1}{R} + j\omega C} = \frac{\frac{1}{R} - j\omega C}{\frac{1}{R^2} + \omega^2 C^2} = \frac{R - j\omega C R^2}{1 + \omega^2 R^2 C^2}$$

We can rearrange this equation, before considering the form of the modulus. Thus, subtracting $R/2$ from each side gives

$$Z - \frac{R}{2} = \frac{R - j\omega C R^2 - \frac{1}{2}R - \frac{1}{2}\omega^2 R^3 C^2}{1 + \omega^2 R^2 C^2}$$

$$= \frac{\frac{1}{2}R(1 - \omega^2 R^2 C^2 - j2\omega CR)}{1 + \omega^2 R^2 C^2}$$

The modulus of $(Z - R/2)$ is thus

$$\left| Z - \frac{R}{2} \right| = \frac{R}{2} \sqrt{\frac{(1 - \omega^2 R^2 C^2)^2 + (2\omega CR)^2}{(1 + \omega^2 R^2 C^2)^2}}$$

$$= \frac{R}{2} \sqrt{\frac{1 - 2\omega^2 R^2 C^2 + \omega^4 R^4 C^4 + 4\omega^2 C^2 R^2}{1 + 2\omega^2 R^2 C^2 + \omega^4 R^4 C^4}}$$

Thus the condition is

$$\left| Z - \frac{R}{2} \right| = \frac{R}{2}$$

This describes a circle of radius $R/2$ centred at $(R/2, 0)$.

The real part of the impedance is

$$\text{real part of } Z = \frac{R}{1 + \omega^2 R^2 C^2}$$

and the imaginary part is

$$\text{imag. part of } Z = \frac{-\omega R^2 C}{1 + \omega^2 R^2 C^2}$$

Thus the argument is

$$\arg Z = \tan^{-1} \left(\frac{-\omega^2 R^2 C}{R} \right) = \tan^{-1}(-\omega^2 RC)$$

For ω varying between 0 and infinity then $\arg Z$ must be between $0°$ and $-90°$.

The modulus and the argument conditions thus indicate that the locus is as shown in figure 6.18. Such a locus indicates that the impedance has a maximum value of R when the argument is zero and as the argument increases from $0°$ to $-90°$ then the impedance diminishes to become zero at $-90°$.

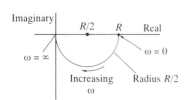

Fig. 6.18 Locus for parallel RC circuit with ω changing

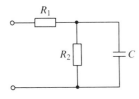

Fig. 6.19 Series RC circuit

Fig. 6.20 Problem 4

Review problems

3   A circuit consists of a resistance in series with a capacitance, with the impedances of the components being as shown in figure 6.19. Sketch the locus of points describing how the impedance of the circuit varies with (a) resistance, (b) angular frequency.

4   For the circuit shown in figure 6.20, determine the locus of points describing how the impedance of the circuit varies with angular frequency.

Further problems

5   Sketch the loci of points satisfying the following conditions:
(a) $\arg(z - 1 - j1) = 90°$,
(b) $30° < \arg(z + 1) < 120°$,
(c) $|z - 2 + j1| = 2$,
(d) $|z - 1 - j1| > 2$,
(e) $|z - 1 - j1| \leq 2$ and $0° < \arg z < 90°$,
(f) $|z + 2 - j2| \leq 1$.

6   A circuit consists of a variable capacitor in series with a resistance R. Sketch the locus describing how the impedance of the circuit varies with the reactance of the capacitor as it varies from 10 Ω to 40 Ω. Hence determine the modulus and the argument of the impedance when $R = 20$ Ω and the capacitive reactance is 30 Ω.

7   A circuit consists of three components in series, a resistance R, an inductance L and a capacitance C. Sketch the locus describing how the impedance of the circuit varies as the angular frequency is increased from zero to infinity.

7 Complex frequency

7.1 Complex frequency

Chapter 3 was a discussion of how phasors, and the use of complex numbers, can be used for the analysis of circuits with steady-state sinusoidal signals, i.e. circuits where there has been sufficient time since the application of the sinusoidal voltage or current for any transients to have died away. This chapter identifies a more general approach that can be used and which applies to exponential signals and exponentially damped sinusoidal signals; it provides a basis for the consideration of a wider range of signals. The method involves what is termed the *complex frequency*.

A sinusoidal voltage with a constant amplitude V_m, and zero phase angle, can be represented by

$$v = V_m \cos \omega t \qquad [1]$$

If, however, we had an amplitude which changed exponentially with time then we could write

$$v = V_m e^{\sigma t} \cos \omega t \qquad [2]$$

where σ is negative when the amplitude decreases with time and positive if it increases with time. This, in fact, is a very useful general equation. The constant amplitude sinusoidal voltage input of equation [1] can be considered to be just a special case of this equation with $\sigma = 0$. If we have just an exponentially changing voltage then we can put $\omega = 0$ and obtain for the voltage $V_m e^{\sigma t}$. We have a steady d.c. voltage of V_m if we put $\sigma = 0$ and $\omega = 0$. Thus, depending on the values of σ and ω, the equation can describe different forms of signal.

Using the equation derived from Euler's formula (chapter 4, equation [8]) the cosine term can be written as

$$\cos \omega t = \tfrac{1}{2}(e^{j\omega t} + e^{-j\omega t})$$

Thus, if we write equation [1] in complex exponential form (see chapter 4) we have,

$$v = \tfrac{1}{2}V_{\mathrm{m}}\, e^{j\omega t} + \tfrac{1}{2}V_{\mathrm{m}}\, e^{-j\omega t} \qquad [3]$$

Equation [2] in complex exponential form is

$$v = \tfrac{1}{2}V_{\mathrm{m}}\, e^{\sigma t} e^{j\omega t} + \tfrac{1}{2}V_{\mathrm{m}}\, e^{\sigma t} e^{-j\omega t}$$

This can be rewritten as

$$v = (\tfrac{1}{2}V_{\mathrm{m}})e^{(\sigma + j\omega)t} + (\tfrac{1}{2}V_{\mathrm{m}})e^{(\sigma - j\omega t)}$$

Hence, since $V_{\mathrm{m}}/2$ is a constant, and if we let

$$s = \sigma + j\omega \qquad [4]$$

then we obtain

$$v = K_1 e^{s_1 t} + K_2 e^{s_2 t} \qquad [5]$$

Equation [5] is now similar to equation [3] with s being like $j\omega$. For this reason s is called the *complex frequency*. Since the power term of st must have no units, then the units of the complex frequency s must be those of $(\text{time})^{-1}$, i.e. second^{-1} or the unit of frequency the hertz (Hz).

There are, in this example, two values of the complex frequency, namely s_1 and s_2. The complex frequency of the first term is $s_1 = \sigma + j\omega$ and that of the second $s_2 = \sigma - j\omega$. The second complex frequency is the complex conjugate of the first complex frequency. A pair of conjugate complex frequencies always occurs if there is a sinusoidal part to the signal.

s has two components, a real part of σ and an imaginary part ω. When the imaginary part is zero then we have an exponentially decaying (or growing) signal, when the real part is zero we have a constant amplitude sinusoidal signal, and when both parts are present we have an exponentially decaying (or growing) sinusoidal signal.

Example

Determine the values of the complex frequencies for the following signals:
(a) $v = 4$, (b) $v = 4 \cos 50t$, (c) $v = 4e^{-5t}$, (d) $v = 4e^{-5t} \cos 50t$.

(a) For such a signal, equation [4] must have the exponential term equal to 1 and so $s = 0 + j0$.

(b) This has only an angular frequency ω and no σ term. Hence we must have $s = 0 + j50$ and the complex conjugate of $0 - j50$.

(c) This has only an exponential term and no sinsusoid. Thus ω is zero and so $s = -5 + j0$. There is only one value since the complex conjugate is $-5 - j0$.

(d) This has an angular frequency ω of 50 and a σ term of -5. Thus $s = -5 + j50$ and the complex conjugate of $-5 - j50$.

Example

Determine the values of the complex frequencies for the signal $v = (1 + e^{-3t}) \cos 50t$.

This signal can be considered to be the sum of two sinusoidal signals, namely $\cos 50t$ and $e^{-3t} \cos 50t$. There will thus be the complex frequencies of $j50$ and $-j50$ for the first signal and for the second signal $-3 + j50$ and $-3 - j50$.

Review problems

1 Determine the values of the complex frequencies for the following signals:
(a) $v = 3$, (b) $v = 3 \cos 50t$, (c) $v = 3e^{-2t}$, (d) $v = 3e^{-2t}\cos 50t$, (e) $v = (4 + e^{-2t}) \cos 50t$.

7.1.1 Complex frequencies for signals with phase angles

In the above discussion the signals had zero phase angle. Now consider the voltage

$$v = V_m \cos (\omega t + \theta)$$

Using the equation derived from Euler's formula (chapter 4, equation [8]), we can write the cosine term as

$$\cos(\omega t + \theta) = \tfrac{1}{2}(e^{j(\omega t+\theta)} + e^{-j(\omega t+\theta)})$$

Thus

$$v = (\tfrac{1}{2}V_m\, e^{j\theta})e^{j\omega t} + (\tfrac{1}{2}V_m\, e^{-j\theta})e^{-j\omega t} \tag{6}$$

The terms in brackets are constants. There is thus the sum of two complex exponentials.

Now consider an exponentially damped sinusoidal voltage

$$v = V_m e^{\sigma t}\cos(\omega t + \theta)$$

Using the expression for the cosine term given above,

$$v = \tfrac{1}{2}V_{\mathrm{m}} \, e^{\sigma t}(e^{j(\omega t+\theta)} + e^{-j(\omega t+\theta)})$$

$$= (\tfrac{1}{2}V_{\mathrm{m}} \, e^{j\theta})e^{(\sigma+j\omega t)} + (\tfrac{1}{2}V_{\mathrm{m}} \, e^{-j\theta})e^{(\sigma-j\omega t)} \qquad [7]$$

The terms in brackets are constants. There is thus the sum of two complex exponentials.

Equation [6] can be considered to be just a version of equation [7] with $\sigma = 0$. Thus, in general we can write

$$v = K_1 e^{s_1 t} + K_2 e^{s_2 t} \qquad [8]$$

where K_1 and K_2 are constants. The complex frequency of the first term is $s_1 = \sigma + j\omega$ and that of the second $s_2 = \sigma - j\omega$. The second complex frequency is the complex conjugate of the first complex frequency.

Example

Determine the values of the complex frequencies for the following:
(a) $v = 5 \cos (100t + 30°)$, (b) $v = 5e^{-4t} \cos (100t + 30°)$.

(a) This has $e^{\sigma t} = 1$ and so $\sigma = 0$. There will thus be two complex frequencies of $0 + j100$ and its complex conjugate $0 - j100$.
(b) This has $e^{\sigma t} = e^{-4t}$ and so $\sigma = -4$. There will be two complex frequencies of $-4 + j100$ and its complex conjugate $-4 - j100$.

Example

Determine the values of the complex frequencies present with the signal $v = (4 - e^{-5t}) \cos (10t + \pi/3)$.

The signal can be considered to be composed of two sinusoidal signals, namely $4 \cos (10t + \pi/3)$ and $-e^{-5t} \cos (10t + \pi/3)$. The complex frequencies for the first signal are $0 + j10$ and $0 - j10$ and for the second $-5 + j10$ and $-5 - j10$. There are thus four complex frequencies.

Example

Determine the general form of the signals which have the following complex frequency components:
(a) 4, −4, (b) −5 + j10, −5 − j10, (c) −5, j10, −j10.

(a) Such a signal, since there is no imaginary components, will have no sinusoidal element. The signal has thus the general form of $Ae^{4t} + Be^{-4t}$, where A and B are constants.
(b) This signal has a conjugate pair of complex frequencies and so is a sinusoidal signal. Since it also has real terms there will also be

an exponential term present. The signal thus has the general form of $Ae^{-5t} \cos(10t + \theta)$, where A and θ are constants.

(c) The -5 complex frequency indicates that there is a component of Ae^{-5t} present. The complex conjugates indicate that there is also a sinusoidal signal of $B \cos 10t$ present. Thus the signal has the general form of $Ae^{-5t} + B \cos 10t$.

Review problems

2 Determine the values of the complex frequencies for the following:
(a) $v = 2 \cos(50t + \pi/4)$, (b) $v = 2e^{3t} \cos(50t - 60°)$,
(c) $v = (5 + e^{-10t}) \cos(4t + 40°)$,
(d) $v = (e^{-10t} + 2e^{-20t}) \cos(50t + \pi/4)$.

3 Determine the general form of the signals which have the following complex frequency components:
(a) 0, 5, -5, (b) $-2 + j10$, (c) 4, j2, $-$ j2, (d) 2, 3 + j2, 3 $-$ j2.

7.2 The *s*-domain

The real part σ of the complex frequency s describes an exponentially varying sinusoid. If it has a negative value then the signal decays as time increases. If positive, the signal increases as time increases. The larger the magnitude of the real part the greater the rate of exponential decay or increase. The imaginary part of s describes the sinusoidal element, being the angular frequency ω. Thus a larger value for the imaginary part means a signal with a higher frequency.

We can plot complex frequencies on an Argand diagram, the real axis being the real part σ of the complex frequency and the imaginary axis being the imaginary part ω. Figure 7.1 shows such a diagram with the forms of signals that give particular complex frequencies being indicated. In the figure the signals have, in the main, been shown for the upper half of the diagram, though it should be recognised that those signals correspond to pairs of conjugate points. Signals in the right half of the figure, i.e. with positive values of σ, increase exponentially with time and eventually will rise to infinity. Thus pairs of conjugate points in the right half represent unstable conditions. Signals in the left half, i.e. with negative values of σ, decay exponentially with time and so are stable signals. Signals for which $\sigma = 0$, but ω does not equal 0, are undamped oscillations. Signals for which ω is 0, but σ is not, are exponentials. Signals for which both σ and ω are 0 are steady d.c. signals.

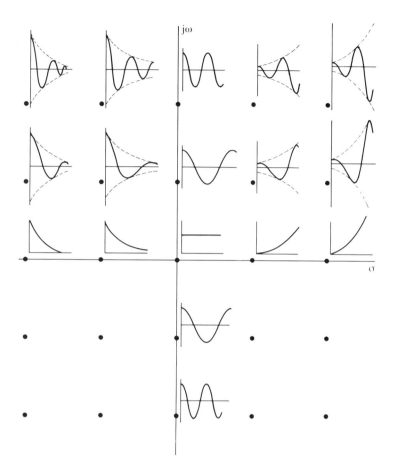

Fig. 7.1 The *s* plane

7.3 Impedance in the *s*-domain

Consider a voltage source of

$$v = V_{\mathrm{m}}\, e^{\sigma t} \cos(\omega t + \theta) \qquad [9]$$

being applied to individual circuit components (figure 7.2). Since this voltage is shown as a function of time it is said to be in the *time-domain*. Euler's formula gives

$$e^{j(\omega t + \theta)} = \cos(\omega t + \theta) + j\,\sin(\omega t + \theta)$$

Thus

$$V_{\mathrm{m}}\, e^{j(\omega t + \theta)} = V_{\mathrm{m}}\,\cos(\omega t + \theta) + j\,V_{\mathrm{m}}\,\sin(\omega t + \theta)$$

Thus we can write *v* as the real part (Re is used in the equation to indicate that it is the real part) of the above equation, i.e.

$$v = \mathrm{Re}\,(V_{\mathrm{m}} e^{\sigma t} e^{j(\omega t + \theta)}) = \mathrm{Re}\,(V_{\mathrm{m}}\, e^{j\theta} e^{st}) \qquad [10]$$

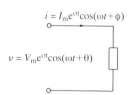

Fig. 7.2 Voltages and currents in the time domain

However, $V_m e^{j\theta}$ is the phasor **V** (see section 4.4) and so equation [10] can be written as

$$v = \text{Re} \, (\mathbf{V}e^{st}) \qquad [11]$$

This represents the voltage signal applied to circuit components. In general, the current in the time-domain can be considered to be of a similar form but with a phase change, i.e.

$$i = I_m \, e^{\sigma t} \cos(\omega t + \phi) \qquad [12]$$

or, using a similar rewriting of the equation as above,

$$i = \text{Re} \, (\mathbf{I}e^{st}) \qquad [13]$$

We can have a general definition of *impedance* in the *s*-domain $Z(s)$ as

$$Z(s) = \frac{v}{i} = \frac{\text{Re} \, (\mathbf{V}e^{st})}{\text{Re} \, (\mathbf{I}e^{st})}$$

Note that $Z(s)$ is *not* Z multiplied by s but the s merely indicates that the impedance is in the *s*-domain.

For a *resistance* R we have the relationship $v = Ri$ and so

$$\text{Re} \, (\mathbf{V}e^{st}) = R \, [\text{Re} \, (\mathbf{I}e^{st})]$$

Thus the impedance in the *s*-domain of a resistance when supplied with a complex frequency is

$$Z(s) = \frac{\mathbf{V}}{\mathbf{I}} = R \qquad [14]$$

For an *inductance* L the voltage v is related to the current i by $v = L \, di/dt$ and so

$$\text{Re} \, (\mathbf{V}e^{st}) = L[\text{Re} \, (s\mathbf{I}e^{st})]$$

and so the impedance in the *s*-domain is

$$Z(s) = \frac{\mathbf{V}}{\mathbf{I}} = sL \qquad [15]$$

For a *capacitance* C the relationship between the voltage v and the current i is $i = C \, dv/dt$ and so

$$\text{Re} \, (\mathbf{I}e^{st}) = C[\text{Re} \, (s\mathbf{V}e^{st})]$$

and so the impedance in the *s*-domain is

$$Z(s) = \frac{\mathbf{V}}{\mathbf{I}} = \frac{1}{sC} \qquad [16]$$

Since the admittance is the reciprocal of inductance then, in the s-domain, the admittance of a resistance is $1/R$, of an inductance $1/sL$ and of a capacitance sC.

Example

Determine the impedances in the s-domain of the following components:
(a) a resistance of 100 Ω,
(b) a capacitance of 10 μF,
(c) an inductance of 0.2 H.

(a) The impedance in the s-domain of a resistance is the resistance (equation [14]) and so is 100 Ω.
(b) The impedance in the s-domain of a capacitance is $1/sC$ (equation [16]) and so is

$$Z(s) = \frac{1}{10 \times 10^{-6} s}\,\Omega$$

(c) The impedance in the s-domain of an inductance is sL (equation [15]) and so is $0.2s$ Ω.

Review problems

4 Determine the impedances in the s-domain of: (a) a resistance of 50 Ω, (b) a capacitance of 2 μF, (c) an inductance of 5 mH.

7.3.1 Series and parallel combinations

$i = \mathrm{Re}(\mathbf{I}e^{st})$

$\mathrm{Re}(\mathbf{V}_1 e^{st})\ \mathrm{Re}(\mathbf{V}_2 e^{st})$

Fig. 7.3 Impedances in series

When impedances are in series (figure 7.3), then the current through each will be the same. Consider a current of $i = \mathrm{Re}\,(\mathbf{I}e^{st})$. For such a current, the potential difference (p.d.) across an impedance will be of the form $\mathrm{Re}\,(\mathbf{V}e^{st})$. The potential difference across the series arrangement will be the sum of the potential differences across each impedance. Hence we have

total p.d. $\mathbf{V}e^{st} = \mathrm{Re}\,(\mathbf{V}_1 e^{st}) + \mathrm{Re}\,(\mathbf{V}_2 e^{st}) + \ldots$

Diving this equation by the current gives

$$\frac{\mathbf{V}e^{st}}{\mathbf{I}e^{st}} = \frac{\mathbf{V}_1 e^{st}}{\mathbf{I}e^{st}} + \frac{\mathbf{V}_2 e^{st}}{\mathbf{I}e^{st}} + \ldots$$

The total impedance will be **V/I** and so the impedance in the *s*-domain is

$$Z(s) = Z_1(s) + Z_2(s) + \dots \qquad [17]$$

Now consider impedances in parallel, as in figure 7.4. The potential difference across each will be the same. Consider the potential difference to be Re $(\mathbf{V}e^{st})$. The current through each impedance will be of the form Re $(\mathbf{I}e^{st})$. Thus, since by Kirchhoff's current law the total current entering the arrangement is the sum of the currents through each impedance,

total current $\mathbf{I}e^{st} = $ Re $(\mathbf{I}_1 e^{st}) + $ Re $(\mathbf{I}_2 e^{st}) + \dots$

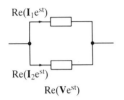

Fig. 7.4 Impedances in parallel

Dividing this equation by the common potential difference across each component gives

$$\frac{\mathbf{I}e^{st}}{\mathbf{V}e^{st}} = \frac{\mathbf{I}_1 e^{st}}{\mathbf{V}e^{st}} + \frac{\mathbf{I}_2 e^{st}}{\mathbf{V}e^{st}} + \dots$$

The total impedance is **V/I** and so the impedance in the *s*-domain is given by

$$\frac{1}{Z(s)} = \frac{1}{Z_1(s)} + \frac{1}{Z_2(s)} + \dots \qquad [18]$$

Thus the same rules apply to series and parallel combinations of impedances in the *s*-domain as apply when considering them with phasor currents and voltages or currents and voltages at some instant of time, i.e. in the frequency or time domains. In a similar way, the techniques for circuit analysis, such as mesh analysis, node analysis, superposition, Thévenin's theorem, Norton's theorem, etc. all apply in the same way in the *s*-domain.

Example

A circuit consists of a resistance R in series with a capacitance C. Determine the impedance of the circuit in the *s*-domain.

The impedance in the *s*-domain of the resistance is R (equation [14]) and that of the capacitor $1/sC$ (equation [16]). Hence the total impedance, using equation [17], is

$$Z(s) = R + \frac{1}{sC}$$

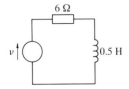

Fig. 7.5 Example

Example

Determine the circuit current for the circuit shown in figure 7.5 when the applied voltage v is $10e^{-2t} \cos(50t + 30°)$ V.

The applied voltage is Re $(\mathbf{V}e^{st})$. The phasor voltage $\mathbf{V}$ is $10\angle 30°$ V. The complex frequency s of the applied voltage is $-2 + j50$. The impedance of the resistance at this complex frequency, i.e. the resistance in the s-domain, is 6 Ω. The impedance of the inductance at this complex frequency is sL and so we have $0.5(-2 + j50) = -1 + j25$ Ω. Since the components are in series, the total impedance is

$$Z(s) = 6 + (-1 + j25) = 5 + j25 \ \Omega$$

Since Re $(\mathbf{I}e^{st})$ = Re $(\mathbf{V}e^{st})/Z(s)$ then, cancelling the e^{st} terms we have $\mathbf{I} = \mathbf{V}/Z(s)$ and so $\mathbf{I}$ is

$$\mathbf{I} = \frac{\mathbf{V}}{Z(s)} = \frac{10\angle 30°}{5 + j25} = \frac{10\angle 30°}{\sqrt{661}\ \angle 78.7°} = 1.96\angle(-48.7°) \ A$$

Thus the circuit current is $0.39e^{-2t} \cos (50t - 48.7°)$ A.

Review problems

5   Determine the impedances in the s-domain of the following:
    (a) a resistance of 10 Ω in series with a capacitance of 0.1 mF,
    (b) a resistance of 2 Ω in series with an inductance of 1 mH,
    (c) a resistance of 20 Ω in parallel with a capacitance of 1 mF,
6   Determine the impedances in the s-domain of the following when subject to a voltage input of $10e^{-4t} \cos 100t$ V:
    (a) a resistance of 10 Ω in series with an inductance of 2 H,
    (b) a resistance of 4 Ω in series with a capacitance of 1 mF,
    (c) a resistance of 1 kΩ in parallel with a capacitance of 1 mF,
    (d) an inductance of 1 H in series with a capacitance of 1 mF.
7   A circuit consists of a resistance R in parallel with an inductance L. Determine the impedance of the circuit in the s-domain.
8   A circuit consists of three components in series, a resistance of 20 Ω, an inductance of 0.2 H and a capacitance of 1 mF. What is the circuit current when a voltage of $6e^{-10t} \cos 100t$ V is applied to the circuit?
9   Determine the impedance in the s-domain of the circuit shown in figure 7.6.

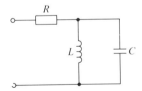

Fig. 7.6 Problem 9

7.4 Circuit analysis in the s-domain

It is often simpler to analyse circuits in terms of their behaviour in the s-domain, rather than the time domain or the frequency-domain. The s-domain describes voltages and currents in terms of sums of complex exponential components involving the complex angular frequency s, in the form e^{st}. The time-domain describes voltages and currents as functions of time. The frequency-domain describes voltages and currents in terms of sums of complex exponential components involving the angular frequency ω, in the form $e^{j\omega t}$. This is just the exponential way of describing phasors.

The advantage of using the s-domain rather than the time-domain is that it is easier to solve equations involving s by means of algebra than the differential equations which occur with the time-domain. The advantage of using the s-domain rather than the frequency-domain is that it is difficult, when terms have been multiplied, to convert an equation from the $j\omega$ form to s form because it is not always possible to know how many j terms were involved. This is because if we have -1 we cannot tell whether no j term was involved or j^2 since $j^2 = -1$. If we have $-j$ we cannot tell whether there is just one j term or whether it is j^3 since $j^3 = -j$. There is another advantage in using the s-domain, the equations so developed can be easily adapted to apply to not only sinusoidal signals, but exponential signals, and signals involving both exponential and sinusoidal changes with time.

To illustrate how a simple circuit can be described in the different domains, consider a circuit consisting of a resistance R in series with an inductance L. In general we can say that the current flowing in the circuit is Re $(\mathbf{I}e^{st})$ when the voltage applied is Re $(\mathbf{V}e^{st})$. The impedance in the s-domain will be

$$Z(s) = R + sL$$

and so the current is

$$\text{Re } (\mathbf{I}e^{st}) = \frac{\text{Re } (\mathbf{V}e^{st})}{R + sL}$$

$$\mathbf{I} = \frac{\mathbf{V}}{R + sL} \qquad [19]$$

In the frequency-domain the current is $\mathbf{I}e^{j\omega t}$ when the applied voltage is $\mathbf{V}e^{j\omega t}$ (see section 4.4). The impedance in this domain, $Z(j\omega)$ will be

$$Z(j\omega) = R + j\omega L$$

and so the current is

$$\mathbf{I}e^{j\omega t} = \frac{\mathbf{V}e^{j\omega t}}{R + j\omega L}$$

$$\mathbf{I} = \frac{\mathbf{V}}{R + j\omega L} \qquad [20]$$

In the time-domain when the current is i the voltage is v. Then the relationship between the voltage and current is given by

$$v = Ri + L\frac{di}{dt} \qquad [21]$$

Equations [19], [20] and [21] thus describe how, in the different domains, the current is related to the voltage.

Consider the circuit involving a resistance in series with an inductance and how, in the s-domain, the general equation for the current, developed earlier as equation [20], can be used with different forms of voltage input. The following illustrates how the equation can be used to solve problems involving a voltage input which is purely exponential and one which is purely sinusoidal. Suppose, for example, we have a voltage of the form $v = V_m e^{\sigma t}$, i.e. just an exponential voltage and no sinusoidal component and so with the complex frequency $s = \sigma + j0$. Thus $\mathbf{V} = V_m \angle 0°$ and so

$$\mathbf{I} = \frac{V_m \angle 0°}{R + \sigma L}$$

The current is then given by

$$\mathrm{Re}\left[\left(\frac{V_m \angle 0°}{R + \sigma L}\right) e^{\sigma t}\right]$$

and so the current in the time domain is

$$i = \left(\frac{V_m}{R + \sigma L}\right) e^{\sigma t} \qquad [22]$$

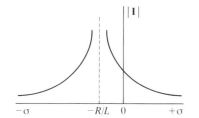

Fig. 7.7 Variation of current with σ

This equation tells us how the current in the circuit depends on σ. When $\sigma = -R/L$, then the denominator of the equation is 0 and so the current is infinite. As $\sigma \to +\infty$ then the current tends to 0. As $\sigma \to -\infty$ the current tends to 0. Figure 7.7 shows how the magnitude of the current varies with σ.

Now consider a voltage of the form $v = V_m \cos \omega t$ applied to the circuit, i.e. there is no exponential term and the complex frequency $s = 0 + j\omega$. Then $\mathbf{V} = V_m \angle 0°$ and so equation [19] can be written as

$$\mathbf{I} = \frac{V_m \angle 0°}{R + j\omega L}$$

Our basic equation in the s-domain has thus been transformed into one in the frequency-domain. Changing the denominator into polar form,

$$\mathbf{I} = \frac{V_m \angle 0°}{\sqrt{R^2 + \omega^2 L^2} \ \angle(\tan^{-1} \omega L/R)}$$

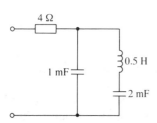

Fig. 7.8 Problem 10

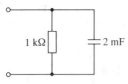

Fig. 7.9 Problem 11

$$= \frac{V_m}{\sqrt{R^2 + \omega^2 L^2}} \angle(- \tan^{-1} \omega L / R)$$

The current in the time-domain is thus

$$i = \frac{V_m}{\sqrt{R^2 + \omega^2 L^2}} \cos\left(\omega t - \tan^{-1} \frac{\omega L}{R}\right) \qquad [23]$$

This equation tells us that the current will be infinite when the denominator is 0, i.e. when $R^2 = -\omega^2 L^2$ or $j\omega = -R/L$.

Review problems

10 For the circuit shown in figure 7.8, determine the impedance in the *s*-domain.

11 For the circuit shown in figure 7.9, determine the impedance in the *s*-domain.

7.4.1 Poles and zeros

In general, the impedance in the *s*-domain can be written in the form

$$Z(s) = K\frac{(s - z_1)(s - z_2) \dots (s - z_n)}{(s - p_1)(s - p_2) \dots (s - p_n)} \qquad [24]$$

When the complex frequency is equal to z_1, z_2, ... z_n then the impedance is zero. Such terms are called *zeros*. The terms p_1, p_2, ... p_n are values of the complex frequency which make the impedance infinite. Such terms are called *poles*.

To illustrate this, consider an impedance

$$Z(s) = \frac{s - 3}{(s - 1)(s - 2)}$$

The impedance is zero when the term $(s - 3) = 0$, i.e. $s = 3$. The impedance is infinite when with the term $(s - 1) = 0$, i.e. $s = 1$, or $(s - 2) = 0$, i.e. $s = 2$.

Example

A circuit has an impedance in the *s*-domain of

$$Z(s) = \frac{0.2s}{1 + 0.002s^2}$$

What value of the complex frequency will make the impedance (a) zero, (b) infinite?

(a) To make the impedance zero the numerator must be zero. Thus we must have $s = 0$.

(b) To make the impedance infinite the denominator must be zero. Thus we must have $0.002s^2 = -1$.

Example

A circuit in the s-domain has poles at $s = -2 + j1$ and $-2 - j1$, and a zero at $s = -3$. Determine an equation for the impedance in the s-domain if the impedance at zero frequency is 12 Ω.

Equation [24] indicates that the impedance will be of the form

$$Z(s) = K\frac{(s+3)}{(s+2-j1)(s+2+j1)} = K\frac{(s+3)}{(s^2+4s+5)}$$

When the complex frequency is zero then

$$Z(s) = K\tfrac{3}{5} = 12$$

Hence $K = 20$ and so

$$Z(s) = 20\frac{(s+3)}{(s^2+4s+5)}$$

Review problems

12 A circuit consists of a capacitance C in series with a resistance R. What value of the complex frequency will make the impedance (a) zero, (b) infinite?

13 A circuit has an impedance in the s-domain of

$$Z(s) = \frac{2s^2+1}{s(2s^2+3)}$$

What value of the complex frequency will make the impedance (a) zero, (b) infinite?

14 A circuit has an inductance of 2 H in parallel with a capacitance of 5 mF. Determine the impedance in the s-domain and hence the value of the complex frequency that will make the impedance (a) zero, (b) infinite?

15 A circuit in the s-domain has poles at $s = -2 + j3$ and $-2 - j3$, and a zero at $s = -2$. Determine an equation for the impedance in the s-domain if the impedance at zero frequency is 4 Ω.

7.5 Transfer function

Often the concern with a circuit is the relationship between the output voltage or current and the input voltage or current. With, for example, an amplifier circuit we are concerned with how much bigger the amplitude of the output is than the input, the term *gain* being often used for this ratio. However, this is only part of the picture, there can also be a phase change produced by the amplifier. Thus if we want to know how the output is related to the input we have to specify how both the amplitude and phase changes. Both the magnitude and phase changes can be incorporated in a single definition relating the output to the input if we consider the input and output in the *s*-domain. This ratio is called the *transfer function*, the symbol $G(s)$ or $H(s)$ generally being used.

$$\text{Transfer function } G(s) = \frac{\text{output } (s)}{\text{input } (s)} \qquad [25]$$

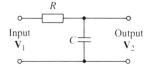

Fig. 7.10 A black box descritpion of a system

We can thus think of a circuit as being a black box which has an input in the *s*-domain and which is converted to an output in the *s*-domain by the transfer function (figure 7.10).

If we are concerned with just a sinusoidal input and output then $s = j\omega$ and so we are concerned with the transfer function in the frequency-domain. We can represent this as $G(j\omega)$. With just a sinusoidal input and output we can represent the input and output by phasors and so $G(j\omega)$ is then the output phasor divided by the input phasor.

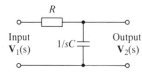

Fig. 7.11 Example

Fig. 7.12 Example

Example

Determine the transfer function for the circuit shown in figure 7.11.

Figure 7.12 shows the circuit of figure 7.11 in the *s*-domain. We have a potential divider circuit and so

$$\frac{V_2(s)}{V_1(s)} = \frac{1/sC}{R + 1/sC}$$

Hence

$$G(s) = \frac{V_2(s)}{V_1(s)} = \frac{1}{sCR + 1}$$

Example

For the circuit described in the previous example, i.e. figure 7.11, determine the output when there is an input of $V_m \cos \omega t$.

The input is sinusoidal so we have $s = j\omega$. The transfer function derived above thus becomes

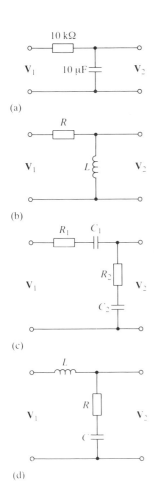

(a)

(b)

(c)

(d)

Fig. 7.13 Problem 16

$$G(j\omega) = \frac{1}{j\omega CR + 1}$$

The output is $G(j\omega)$ multiplied by the input, namely $V_m \angle 0°$. Thus the output is

$$\mathbf{V}_2 = \frac{1}{j\omega CR + 1} V_m \angle 0°$$

$$= \frac{1}{1 + j\omega CR} \times \frac{1 - j\omega CR}{1 - j\omega CR} V_m \angle 0°$$

$$= \frac{1 - j\omega CR}{1 + \omega^2 C^2 R^2} V_m \angle 0°$$

The magnitude of the output is therefore

$$|\mathbf{V}_2| = \frac{V_m}{\sqrt{1 + \omega^2 C^2 R^2}}$$

and the phase is

$$\phi = -\tan^{-1} \omega CR$$

Review problems

16 Determine the transfer functions for the circuits shown in figure 7.13.
17 Determine the output for the circuit shown in figure 7.13(a) when the input is 4 cos 500*t* V.

7.6 Frequency response of circuits

The frequency response of a circuit is the description of how the phasor representing the output varies in amplitude and phase angle as the sinusoidal input is varied in frequency, i.e. how the transfer function varies as the sinusoidal input is varied in frequency. Such responses can be displayed by means of a *Bode plot*. This consists of two graphs plotted on logarithmic or semilogarithmic axes, one showing how the modulus, i.e. the magnitude or amplitude, varies with frequency and the other how the argument, i.e. the phase angle, varies with frequency.

Another way of displaying the frequency response is by means of a locus plot (see chapter 6). This is a plot showing how the real and imaginary parts of the transfer function or output phasor vary with frequency. This is a form often referred to as a

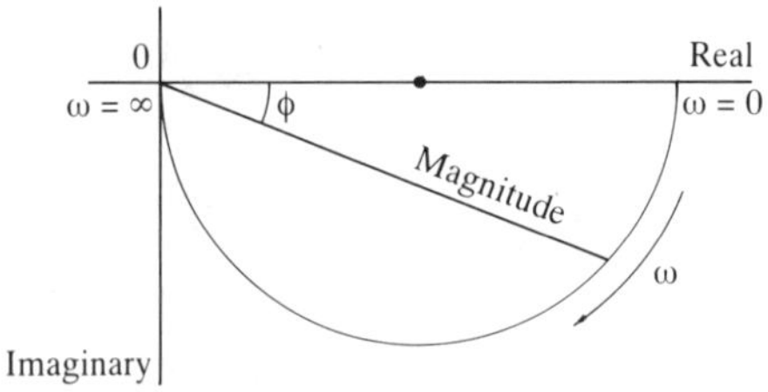

Fig. 7.14 A Nyquist diagram

Nyquist diagram. Figure 7.14 shows the form such a diagram can take for a circuit consisting of capacitance in series with resistance and the output is the voltage across the capacitance. The length of the line originating at the origin represents the magnitude and the angle of the line to the axis the phase angle. The plot indicates that when the frequency is zero that the transfer function or output has zero phase angle and the magnitude a maximum value. Then as the frequency is increased the phase angle increases and the magnitude diminishes.

Example

Sketch the Bode plot for the circuit given in figure 7.11 and considered in the previous two examples.

For the Bode plot we have a sinusoidal input to the circuit, as in the previous example. Thus the magnitude of the transfer function is

$$|G(j\omega)| = \frac{1}{\sqrt{1 + \omega^2 C^2 R^2}}$$

and the angle is

$$\phi = -\tan^{-1} \omega CR$$

The magnitude of the transfer function is plotted against ω and the angle is also plotted against ω. For $\omega = 0$ we have the magnitude equal to 1 and the angle to $0°$. For $\omega = \infty$ we have the magnitude equal to 0 and the angle to $-90°$. When $\omega RC = 1$ we have the magnitude equal to $1/\sqrt{2}$ and the angle to $-45°$. Figure 7.15 shows the resulting Bode plot.

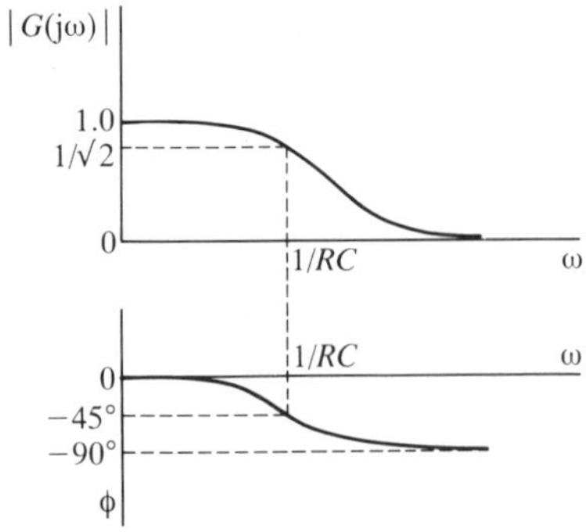

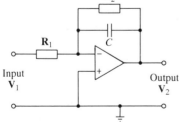

Fig. 7.15 Example

Fig. 7.16 Problem 19

Review problems

18 Sketch the Bode plot for the circuit shown in figure 7.13(b).
19 Sketch the Bode plot for the operational amplifier circuit shown in figure 7.16 if the amplifier can be assumed to be ideal with the transfer function being −(feedback impedance)/(input impedance).

Further problems

20 What are the complex frequencies for the following signals:
(a) $v = 10$, (b) $v = 10 \cos 100t$, (c) $v = 10e^{-4t}$, (d) $v = 10e^{4t}$,
(e) $v = 10e^{-4t} \cos 100t$, (f) $v = 10 \cos (100t + \pi/2)$,
(g) $v = 10e^{-4t} \cos(100t - 20°)$, (h) $v = (1 + e^{-10t}) \cos 100t$,
(i) $v = (2e^{-2t} + 5e^{-3t}) \cos (100t + \pi/3)$.

21 Determine the general form of the signals which have the following complex frequency components:
(a) 0, 3, -3, (b) $2 + j2$, $2 - j2$, (c) 0, $-4 + j2$, $-4 - j2$,
(d) -5, $j6$, $-j6$

22 Determine the impedances in the s-domain of the following:
(a) a resistance of 1 kΩ,
(b) a capacitance of 8 μF,
(c) an inductance of 20 mH,
(d) an inductance L in series with a capacitance C,
(e) a resistance R in parallel with a capacitance C,
(f) a resistance of 10 Ω in series with an inductance of 0.1 H,
(g) a resistance of 10 Ω in parallel with an inductance of 0.1 H,
(h) a resistance of 10 Ω in parallel with a capacitance of 1 mF.

23 A circuit consists of a resistance of 20 Ω in series with an inductance of 4 H. How will the current in the circuit vary with time when a voltage of $8e^{-5t} \cos 10t$ V is applied?

24 A circuit consists of an inductance of 1 H in parallel with a capacitance of 2 mF. What is the impedance in the s-domain and what values of complex frequency will make the impedance (a) zero, (b) infinite?

25 A circuit has an impedance in the s-domain given by

$$Z(s) = \frac{(s+1+j3)(s+1-j3)}{(s+j1)(s-j1)}$$

What values of complex frequency will make the impedance (a) zero, (b) infinite?

26 A circuit has an impedance function in the s-domain which has poles at $-1 + j4$ and $-1 - j4$, and a zero at -2. What will be the form of the impedance in the s-domain?

27 A circuit has an impedance function in the s-domain which has poles at $-2 + j4$ and $-2 - j4$, and a zeros at -3 and -5. Determine an equation for the impedance in the s-domain if the impedance at zero frequency is 15 Ω.

28 Determine the transfer functions for the circuits shown in figure 7.17.

29 Many circuits have transfer functions of the form

$$H(s) = \frac{1}{1 + \tau s}$$

where τ is the time constant of the circuit. Sketch the Bode plot for such circuits.

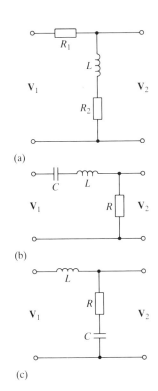

(a)

(b)

(c)

Fig. 7.17 Problem 28

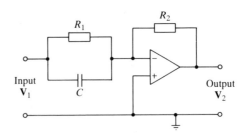

Fig. 7.18 Problem 30

30 Determine the transfer function for the operational amplifier circuit shown in figure 7.18 if the amplifier can be assumed to be ideal with the transfer function being −(feedback impedance)/(input impedance).

Answers

Chapter 1

1 (a) $\pm j5$, (b) $-2 \pm j3$, (c) $0.5 \pm j1.1$

2 (a) $-3 - j2$, (b) $5 + j3$, (c) $1 + j0.5$

3 See figure A.1

4 (a) 3.6, 30.5°, (b) 1.4, 135°, (c) 1.4, 315°

5 (a) $5.0\angle53.1°$, (b) $4\angle270°$, (c) $3\angle0°$, (d) $9.45\angle122°$,
   (e) $9.45\angle-58°$

6 (a) $3.5 + j2.0$, (b) -6, (c) $+2$, (d) $-5.7 - j4.0$

7 (a) $10\angle60°$ N, $5.0 + j8,7$ N, (b) $1\angle120°$ kN, $-0.5 + j0.87$ kN,
   (c) $5\angle45°$ m/s, $3.5 + j3.5$ m/s, (d) $10\angle140°$ m/s,
   $-7.7 + j6.4$ m/s

8 (a) $\pm j4$, (b) $-0.5 \pm j1.5$, (c) ±3, (d) $0.25 \pm j0.66$,
   (e) $-1 \pm j2.24$

9 (a) $-2 - j3$, (b) $5 - j6$, (c) $-7 + j7$

10 (a) $2\angle108.4°$, (b) $2\angle251.6°$, (c) $2\angle71.6°$, (d) $1\angle0°$,
   (e) $3\angle90°$

11 (a) 10, (b) $j10$, (c) $-j10$, (d) $8.7 + j5.0$, (e) $-8.7 + j5.0$

12 (a) $100\angle30°$ N, $86.6 + j50$ N, (b) $100\angle90°$ N, $j100$ N,
   (c) $12\angle30°$ m/s, $10.3 + j6.0$ m/s, (d) $4\angle180°$ m/s, $-4 + j0$ m/s,
   (e) $10\angle45°$ m/s^2, $7.07 + j7.07$ m/s^2, (f) $5\angle230°$ m/s^2,
   $-3.21 - j3.83$ m/s^2

Chapter 2

1 $a = 2, b = 7$

2 (a) $3 - j6$, (b) $1 + j2$

3 (a) $4 + j7, -2 + j3$, (b) $4 + j2, 2 - j10$, (c) $1 + j7, -5 - j1$,
   (d) $2 + j2, 4 + j6$

4 (a) $12 + j3$, (b) $-21 + j1$, (c) $-8 + j2$, (d) $9 + j8$

5 (a) $10\angle 70°$, (b) $15\angle 160°$, (c) $12\angle 10°$, (d) $6\angle(-20°)$

6 (a) 25, (b) 29, (c) 40, (d) 41

7 (a) $0.71 - j0.82$, (b) $1 + j1$, (c) $0.5 - j1.5$, (d) $0.21 - j0.14$

8 (a) $2\angle 20°$, (b) $2\angle 70°$, (c) $0.67\angle(-60°)$, (d) $0.8\angle 60°$

9 (a) 4, (b) $-2 + j4$, (c) $2 - j4$, (d) $7 + j4$, (e) $-0.2 - j1.6$,
   (f) $12 - j8$, (g) $-2.52 - j0.64$

10 (a) $0.24\angle(-20°)$, (b) $16\angle 40°$, (c) $0.0576\angle(-40°)$

11 $1.6\angle(-30°)$

12 $j0.18$

13 (a) $5 + j5$, $-1 + j1$, (b) $6 - j3$, $2 + j7$, (c) $4 + j2$, $0 - j8$,
   (d) $-1 + j2$, $-5 - j8$, (e) $-1 + j4$, $5 - j6$, (f) $-5 + j8$, $-3 + j4$

14 (a) $23 + j14$, (b) $8 - j6$, (c) $-2 - j16$, (d) $10 + j20$,
   (e) $-10 + j60$, (f) $0 - j13$

15 (a) $60\angle 55°$, (b) $12\angle 130°$, (c) $20\angle 10°$, (d) $8\angle(-110°)$,
   (e) $60\angle 0°$, (f) $20\angle 80°$

16 (a) 26, (b) 26, (c) 13, (d) 41

17 (a) $0.44 + j0.049$, (b) $0.12 + j0.53$, (c) $-0.1 - j0.7$, (d) -1.5

18 (a) $0.5\angle 0°$, (b) $3\angle 60°$, (c) $0.5\angle(-80°)$, (d) $2\angle 70°$

19 (a) $5 + j2$, (b) $-10 + j10$, (c) $-0.23 + j0.85$, (d) $0.35 - j0.95$,
   (e) $19 - j4$

20 $0.9 + j1.3$

21 (a) $3 - j1$, (b) $1 - j1$, (c) $8 + j1$, (d) $-0.6 + j0.8$, (e) $4 - j2$,
   (f) $1 + j1$, (g) $2.6 + j2.2$

22 (a) $0.2\angle(-30°)$, (b) $25\angle 60°$, (c) $0.04\angle(-60°)$

23 (a) 2, $2\angle 0°$, (b) $0 + j1$, $1\angle 90°$, (c) $j2$, $2\angle 90°$

24 (a) $3 - j1$, $3.2\angle-18.4°$, (b) $j1$, $1\angle 90°$, (c) $1 + j3$, $3.2\angle 71.6°$

25 $2.22\angle 248°$

26 (a) $4.71\angle(-5.8°)$, (b) $7.60\angle 35°$

27 (a) $-j/\omega C$, $(1/\omega C)\angle(-90°)$,

   (b) $R - j/\omega C$, $\left(\sqrt{R^2 + 1/\omega^2 C^2}\right)\angle \tan^{-1}(-1/\omega CR)$,

   (c) $j(\omega L - 1/\omega C)$, $(\omega L - 1/\omega C)\angle 90°$,

   (d) $R + j(\omega L - 1/\omega C)$,
   $\left(\sqrt{R^2 + (\omega L - 1/\omega C)^2}\right)\tan^{-1}(\omega L - 1/\omega C)/R$

Chapter 3

1   (a) $4\sin(100\pi t + 60°)$ V, $2.8\angle 60°$ V,

    (b) $0.5\sin(100\pi t + 20°)$ A, $0.35\angle 20°$ A,

    (c) $10\sin(2000\pi t + 90°)$ V, $7.07\angle 90°$ V,

    (d) $2.83\sin(100\pi t + 60°)$ A, $2\angle 60°$ A.

2   (a) $14.1\angle 40°$ mA, (b) $7.07\angle(-30°)$ V, (c) $14.3\angle(-120°)$ mA,

    (d) $3.53\angle 0°$ V

3   (a) $4.2\sin(\omega t + 20°)$ A, (b) $5.65\sin(\omega t - 60°)$ V,

    (c) $1.7\sin(\omega t + 90°)$ A, (d) $141\sin(\omega t - 30°)$ V,

    (e) $70.7\sin\omega t$ mA

4   $4.5\sin(\omega t + 63.4°)$ V

5   $5\sin(\omega t + 53.1°)$ mA

6   $44.7\sin(\omega t - 63.4°)$ mA

7   $10.8\sin(\omega t - 68.2°)$ V

8   (a) $5\angle 53.1°$, (b) $5\angle(-36.9°)$, (c) $2.8\angle 225°$, (d) $1.7\angle 116.6°$

9   (a) $6 + j2$, (b) $-2.5 + j4.3$, (c) $1.7 - j1.0$, (d) $j4$

10 $6.1\angle 25.3°$ V

11 $14\angle 30.4°$ mA

12 $25.4\angle 209.6°$ V

13 $5.04\angle 79.1°$ A

14 $10\angle 53.1°$ V

15 $8.06\angle 7.1°$ A

16 $7.07\angle 8.1°$ V

17 (a) $5\angle 60°$ Ω, (b) $4\angle 90°$ Ω, (c) $1.85\angle(-37.2°)$ Ω,

    (d) $4\angle(-30°)$ Ω

18 (a) $41.2\angle 74°$ V, (b) $2\angle 120°$ V, (c) $11.2\angle(-33.4°)$ V,

    (d) $15\angle 30°$ V

19 (a) $10 + j4$ Ω, $5 + j5$ Ω, (b) $5 + j6$ Ω, $1.41 + j1.51$ Ω,

    (c) $2.5 - j2.3$ Ω, $0.22 + j146$ Ω, (d) $5.2 + j1$ Ω, 2.30 Ω

20 (a) $- j0.063$ Ω, (b) $j12\,566$ Ω, (c) 100 Ω

21 (a) $10 + j6.28$ Ω, (b) $10 + j62.8$ Ω, (c) $1 - j1.59$ kΩ,

    (d) $1 - j0.159$ kΩ

22 $500\angle 120°$ V

23 $0.36\angle 135°$ mA

24 $7.9\angle 45°$ A

25 (a) $2.88 + j3.84$ Ω, (b) $2.25\angle(-53.1°)$ A

26 $0.14\angle 45°$ A

27 $17.9\angle(-63.4°)$ V

28 2664 VA, 1838 W, 0.69 lag

29 22.4 VA, 20.0 W, 0.89 lag

30 $12 + j6$ VA, 12 W, 16 VAr

31 $866 + j500$ VA

32 (a) $3.5\angle 0°$ V, (b) $7.1\angle 90°$ V, (c) $14.1\angle 60°$ mA,
    (d) $1.4\angle 120°$ A , (e) $3.5\angle(-30°)$ V, (f) $14.1\angle(-90°)$ mA

33 (a) $28.2 \sin(\omega t + 30°)$ V, (b) $212.1 \sin(\omega t + 120°)$ mA,
    (c) $4.2 \sin(\omega t - 20°)$ V, (d) $0.57 \sin \omega t$ A

34 (a) $3.6\angle(-56.3°)$, (b) $5.8\angle 121°$, (c) $9.4\angle(-60°)$,
    (d) $5.8\angle 59°$

35 (a) $-j3$, (b) $4.7 + j1.7$, (c) $1.0 - j1.7$, (d) $-3.3 + j2.3$

36 $7.03\angle 12.3°$ V

37 $11.1\angle 8.9°$ mA

38 $6\angle 116°$ V

39 $4.31\angle(-34.8°)$ A

40 $11.4\angle 37.9°$ V

41 $9.22\angle 40.6°$ A

42 $8.6\angle(-35.5°)$ V

43 $1 - j2$ A

44 $-4 + j14$ V

45 (a) $240\angle 50°$ Ω, (b) $10\angle 60°$ Ω, (c) $5\angle 120°$ Ω, (d) $100\angle 90°$ Ω

46 (a) $50\angle 160°$ V, (b) $8\angle(-10°)$ V, (c) $10.3\angle(-24.1°)$ V,
    (d) $0.5\angle 120°$ V

47 (a) $12 + j5$ Ω, (b) $10 + j3$ Ω, (c) $23.8 + j15.5$ Ω,
    (d) $140 - j10$ Ω

48 (a) $20 + j40$ Ω, (b) $4.73 + j0.88$ Ω, (c) $0.77 - j3.85$ Ω,
    (d) $j13.3$ Ω

49 (a) $32.9\angle 13°$ Ω, (b) $0.37\angle(-13°)$ A,
    (c) $7.3\angle 17°$ V, $5.5\angle(-23°)$ V

50 (a) 1 kΩ, (b) $j62.8$ Ω, (c) $-j0.16$ Ω, (d) $1000 + j62.8$ Ω,
    (e) $1000 - j0.16$ Ω, (f) $1000 + j62.6$ Ω

51 $222\angle(-64.4°)$ V, $96\angle 0°$ V, $200\angle(-90°)$ V

52 $250\angle 30°$ V

53 $1.44\angle 110°$ mA

54 $63.3 \sqrt{2} \sin(1000t - 18.4°)$ V

55 $5.38 \sqrt{2} \sin(400t + 79.3°)$ A

56 $4\angle(-127°)$ mA

57 $0.54\angle(-17.1°)$ A

58 $1.06\angle(- 45°)$ A

59 2644 VA, 2620 W, 0.99 lag

60 354 VA, 250 W, 0.71 lag

61 20.8 + j12 VA

62 3.83 + j3.21 VA

63 4.2∠(−56.3°) A, 34.6 + j51.9 VA

Chapter 4

1 (a) 1.10517, (b) 0.99500 + j0.09983, (c) 0.9950 − j0.09983

2 (a) 5∠π/2, j5, (b) 2∠(−2π/3), − 1 − j1.73

3 (a) je^2, (b) e^4(0.87 − j0.5), (c) e$^\pi$(cos 1 − j sin 1)

4 (a) 3e$^{j\pi/2}$, (b) $\sqrt{5}$ e$^{-j\pi/4}$, (c) $\sqrt{10}$ e$^{j\,\tan^{-1}(1/3)}$

5 (a) $\dfrac{e^{j4} - e^{-j4}}{2j}$, (b) $\dfrac{e^{-2} + e^2}{2}$, (c) $\dfrac{e^{-3} - e^3}{2j}$

6 As problem

7 $\frac{3}{4}$ sin θ − $\frac{1}{4}$ sin 3θ

8 $\frac{1}{16}$ cos 5θ + $\frac{5}{16}$ cos 3θ + $\frac{5}{8}$ cos θ

9 (a) −j sin4, (b) cosh π/4, (c) j tan π/3

10 As problem

11 As problem

12 (a) ln 1 + j2nπ, (b) ln $\sqrt{2}$ + j(2n + 1)π,
    (c) ln $\sqrt{5}$ + j(1.11 + 2nπ)

13 (a) Imag. part 2e$^{j(100t+45°)}$, (b) Real part 12e$^{j(500t−60°)}$,
    (c) Imag. part 5e^{j300t}

14 (a) 4∠π/4, 2.83 + j2.83, (b) 2∠π/3, 1 + j1.73,
    (c) 3∠(−π/6), 2.60 − j1.5

15 (a) −1, (b) −j2, (c) 4(0.86 − j0.5), (d) e^2(cos 2 − j sin 2)

16 (a) 5.66e$^{j\pi/4}$, (b) 5e$^{j\pi/2}$, (c) 5.39e$^{j0.38\pi}$, (d) 4e^0

17 As problem

18 $\frac{5}{16}$ − $\frac{15}{32}$ cos 2θ + $\frac{3}{16}$ cos 4θ − $\frac{1}{32}$ cos 6θ

19 $\frac{35}{64}$ sin θ − $\frac{21}{64}$ sin 3θ + $\frac{7}{64}$ sin 5θ − $\frac{1}{64}$ sin 7θ

20 (a) j sinh π/4, (b) j sin π/4, (c) cosh π/3, (d) j tanh π/3

21 3.17 + j1.96

22 −6.55 − j7.62

23 (a) ln 2 + j(π/2 + 2nπ), (b) ln 9 + j(π + 2nπ),
    (c) ln $\sqrt{20}$ + j(1.11 + 2nπ)

24 (a) $\sqrt{5}$ e$^{j63.4°}$, (b) $\sqrt{20}$ e$^{-j26.6°}$, (c) 25e$^{-j30°}$, (d) 10e$^{j40°}$

25 As problem

26 ± 2.63

Chapter 5

1 (a) $1\angle 0°$, $1\angle 60°$, $1\angle 120°$, $1\angle 180°$, $1\angle 240°$, $1\angle 300°$, $1\angle 360°$,
   1, $0.5 + j0.87$, $-0.5 + j0.87$, -1, $-0.5 - j0.87$, $0.5 - j0.87$,
   (b) $0.84\angle 67.5°$, $0.84\angle 247.5°$, $0.32 + j0.78$, $-0.32 - j0.78$,
   (c) $1\angle 30°$, $1\angle 150°$, $1\angle 270°$, $0.87 + j0.5$, $-0.87 + j0.5$, $-j1$,
   (d) $1.12\angle 15°$, $1.12\angle 135°$, $1.12\angle 255°$, $1.08 + j0.29$,
   $-0.79 + j0.79$, $-0.29 - j0.76$

2 (a) $-2 + j2$, (b) $-117 + j44$, (c) $-117 + j44$, (d) $117 + j44$

3 $2, j2, -2, -j2$

4 $1, 0.31 + j0.95, 0.31 - j0.95, 0.81 + j0.59, 0.81 - j0.59$

5 (a) $0.56 + j0.18$, $-0.43 + j0.39$, $-0.12 - j0.57$,
   (b) $0.056 - j0.422$, $0.337 + j0.259$, $-0.393 + j0.163$,
   (c) $1 + j1$, $-1.37 + j0.37$, $-0.37 - j1.37$

6 (a) $2\angle 40°$, $2\angle 160°$, $2\angle 280°$, (b) $2.11\angle 10°$, $2.11\angle 100°$,
   $2.11\angle 190°$, $2.11\angle 280°$, (c) $1.64\angle 60°$, $1.64\angle 132°$, $1.64\angle 204°$,
   $1.64\angle 276°$, $1.64\angle 348°$

7 $\cos^4\theta - 6\cos^2\theta \sin^2\theta + \sin^4\theta$

8 $5\cos^4\theta \sin\theta - 10\cos^2\theta \sin^3\theta + \sin^5\theta$

9 $\cos^8\theta - 28\cos^6\theta \sin^2\theta + 70\cos^4\theta \sin^4\theta - 28\cos^2\theta \sin^6\theta + \sin^8\theta$

10 (a) $2, -1 + j1.73, -1 - 1.73$, (b) $1.07 - j0.21, 0.21 + j1.07$,
    $-1.07 + j0.21, -0.21 - j1.07$, (c) $-0.61 + j1.69, 1.16 - j1.37$,
    $1.77 - j0.32$, (d) $0.71 + j0.71, -0.71 - j0.71$

11 (a) $-512 + j887$, (b) $0.016 - j0.088$, (c) $-237 + j3116$,
    (d) $2 + j2$

12 $1.41 + j1.41, -1.41 + j1.41, -1.41 - j1.41, 1.41 - j1.41$

13 $0.71 + j0.71, -0.71 - j0.71$

14 (a) $3\angle 30°$, $3\angle 210°$, (b) $5\angle 30°$, $5\angle 150°$, $5\angle 270°$, (c) $3\angle 20°$,
    $3\angle 110°$, $3\angle 200°$, $3\angle 290°$, (d) $1.58\angle 15°$, $1.58\angle 51°$,
    $1.58\angle 87°$, $1.58\angle 123°$, $1.58\angle 159°$, $1.58\angle 195°$, $1.58\angle 231°$,
    $1.58\angle 267°$, $1.58\angle 303°$, $1.58\angle 339°$

15 $4\cos^3\theta \sin\theta - 4\cos\theta \sin^3\theta$

16 After using $\sin^2\theta = 1 - \cos^2\theta$, then $32\cos^6\theta - 48\cos^4\theta + 18\cos^2\theta - 1$

17 After using $\cos^2\theta = 1 - \sin^2\theta$, then $16\sin^4\theta - 20\sin^2\theta + 5$

18 $\dfrac{5\tan\theta - 10\tan^3\theta + \tan^5\theta}{1 - 10\tan^2\theta + 5\tan^4\theta}$

Chapter 6

1 See figure A.2
2 See figure A.3

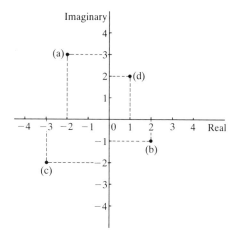

Fig. A.1 Chapter 1 Problem 3

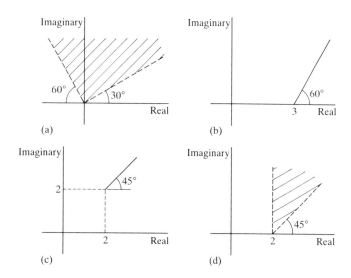

Fig. A.2 Chapter 6 Problem 1

3   See figure A.4

4   See figure A.5

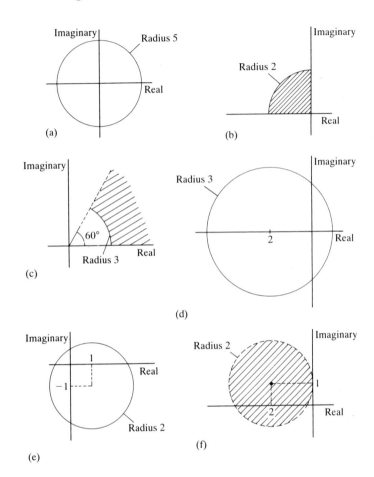

Fig. A.3 Chapter 6 Problem 2

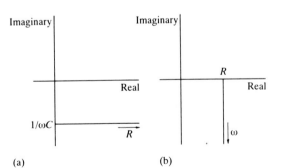

Fig. A.4 Chapter 6 Problem 3

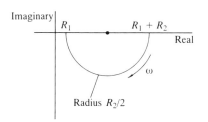

Fig. A.5 Chapter 6 Problem 4

5 See figure A.6
6 See figure A.7
7 See figure A.8

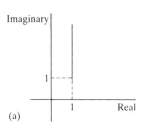

(a)

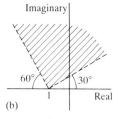

(b)

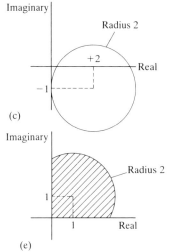

(c)

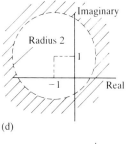

(d)

(e)

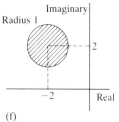

(f)

Fig. A.6 Chapter 6 Problem 5

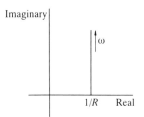

Fig. A.7 Chapter 6 Problem 6

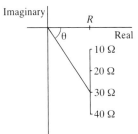

Fig. A.8 Chapter 6 Problem 7

Chapter 7

1 (a) $0 + j0$, (b) $0 + j50$, $0 - j50$, (c) $-2 + j0$, (d) $-2 + j50$, $-2 - j50$, (e) $j50$, $-j50$, $-2 + j50$, $-2 - j50$

2 (a) $j50$, $-j50$, (b) $3 + j50$, $3 - j50$, (c) $j4$, $-j4$, $-10 + j4$, $-10 - j4$, (d) $-10 + j50$, $-10 - j50$, $-20 + j50$, $-20 - j50$

3 (a) $A + Be^{5t} + Ce^{-5t}$, (b) $Ae^{-2t} \cos(10t + \theta)$,
   (c) $Ae^{4t} + B \cos(2t + \theta)$, (d) $Ae^{2t} + Be^{3t} \cos(2t + \theta)$

4 (a) $50\ \Omega$, (b) $1/(2 \times 10^{-6}s)\ \Omega$, (c) $0.005s\ \Omega$

5 (a) $10 + \dfrac{1}{10^{-4}s}\ \Omega$, (b) $2 \times 10^{-3}\ s\ \Omega$, (c) $\dfrac{1}{0.05 + 0.001s}\ \Omega$

6 (a) $2 + j100\ \Omega$, (b) $3.6 - j9.98\ \Omega$, (c) $1/(-0.003 + j0.1)\ \Omega$,
   (d) $-4.4 - j90\ \Omega$

7 $\dfrac{sLR}{R + sL}$

8 $0.30\angle(-30.7°)$ A

9 $\dfrac{s^2RLC + sL + R}{s^2LC + 1}$

10 $\dfrac{0.001s^2 + 1}{0.001s(0.001s + 3)}$

11 $\dfrac{1000}{1 + 2s}$

12 (a) $-1/RC$, (b) 0

13 (a) $s = j\sqrt{0.5}$ Hz, (b) $s = 0$, $s = j\sqrt{1.5}$ Hz

14 (a) $s = 0$, (b) $s = j10$ Hz

15 $26\dfrac{(s + 2)}{(s^2 + 4s + 13)}$

16 (a) $\dfrac{1}{0.1s + 1}$, (b) $\dfrac{sL}{R + sL}$, (c) $\dfrac{R_2 + 1/sC_2}{R_1 + R_2 + (1/sC_1) + (1/sC_2)}$,
   (d) $\dfrac{R}{L}\dfrac{s + 1/RC}{s^2 + s(R/L) + (1/LC)}$

17 $3.98 \cos(500t - 5.7°)$ V

18 As in figure 7.15 with $\phi = -45°$ at $\omega = R/L$

19 As in figure 7.15 with $\phi = -45°$ at $\omega = 1/R_2C$ and all
   magnitude values multiplied by R_2/R_1

20 (a) $0 + j0$, (b) $0 + j100$, $0 - j100$, (c) $-4 + j0$, (d) $4 + j0$,
   (e) $-4 + j100$, $-4 - j100$, (f) $0 + j100$, $0 - j100$,
   (g) $-4 + j100$, $-4 - j100$, (h) $j100$, $-j100$, $-10 + j100$,
   $-10 - j100$, (i) $-2 + j100$, $-2 - j100$, $-3 + j100$, $-3 - j100$

21 (a) $A + Be^{3t} + Ce^{-3t}$, (b) $Ae^{2t} \cos(2t + \theta)$,
   (c) $A + Be^{-4t} \cos(2t + \theta)$, (d) $Ae^{-5t} + B \cos(6t + \theta)$

22 (a) $1\ k\Omega$, (b) $\dfrac{1}{8 \times 10^{-6}s}\ \Omega$, (c) $20s\ m\Omega$, (d) $sL + \dfrac{1}{sC}$,

(e) $\dfrac{R}{1+sCR}$, (f) $10+0.1s$ Ω, (g) $\dfrac{10s}{s+10}$ Ω, (h) $\dfrac{1}{0.1+0.01s}$ Ω

23 $0.20\angle(-90°)$ A

24 $Z(s)=\dfrac{0.002s^2+1}{0.002s}$ Ω, (a) $s=j\sqrt{500}$ Hz, (b) $s=0$

25 (a) $-(1+j3)$, $-(1-j3)$, (b) $-j1$, $j1$

26 $Z(s)=K\dfrac{(s+2)}{(s+1-j4)(s+1+j4)}$

27 $20\dfrac{(s+3)(s+5)}{(s^2+4s+20)}$

28 (a) $\dfrac{R_2+sL}{R_1+R_2+sL}$, (b) $\dfrac{R}{R+sL+1/sC}$, (c) $\dfrac{sRC+1}{s^2LC+sRC+1}$

29 As in figure 7.15 with $\phi=-45°$ at $1/\tau$

30 $-R_2C\left(s+\dfrac{1}{R_1C}\right)$

Index